P9-APG-333

Nanoethics

Think Now

Think Now is a brand new series of stimulating and accessible books examining key contemporary social issues from a philosophical perspective. Written by experts in philosophy, these books offer sophisticated and provocative yet engaging writing on political and cultural themes of genuine concern to the educated reader.

Available now:

The Ethics of Climate Change, James Garvey
War and Ethics, Nicholas Fotion
Terrorism, Nicholas Fotion, Boris Kashnikov and Joanne K. Lekea
Personal Responsibility, Alexander Brown

Forthcoming:

Beyond Animal Rights, Tony Milligan
Digital Justice, Justine Johnstone
Identity Crisis, Jeremy Stangroom
Just Warriors, Inc., Deane-Peter Baker
The Ethics of Metropolitan Growth, Robert Kirkman
The Ethics of Trade and Aid, Christopher D. Wraight

Series Editors:

James Garvey is Secretary of The Royal Institute of Philosophy and author of *The Twenty Greatest Philosophy Books* (Continuum).
Jeremy Stangroom is co-editor, with Julian Baggini, of *The Philosophers' Magazine* and co-author of *Why Truth Matters, What Philosophers Think* and *Great Thinkers A–Z* (all Continuum).

Nanoethics

Big Ethical Issues with Small Technology

Dónal P. O'Mathúna

continuum

Continuum

Continuum International Publishing Group
The Tower Building
11 York Road
London SE1 7NX

80 Maiden Lane
Suite 704
New York NY 10038

www.continuumbooks.com

British Library Cataloguing-in-Publication Data
A catalogue record for this book is available from the British Library.

ISBN: HB: 978-1-8470-6394-6
PB: 978-1-8470-6395-3

Library of Congress Cataloging-in-Publication Data
A catalog record for this book is available from the Library of Congress.

Typeset by Servis Filmsetting Ltd, Stockport, Cheshire
Printed and bound in Great Britain by the MPG Books Group, UK

For Catrina, Conor and Peter
It's your future

Contents

Preface

Nanotechnology is the new buzz word in many parts of academia, industry and government. Some are claiming it is the key to the next industrial revolution. It is poised to change every aspect of human life, and maybe even humanity itself. Yet most people seem completely unaware of this impending revolution. When I tell people that I'm interested in the ethics of nanotechnology, they look at me blankly or inquisitively. When I describe some of nanotechnology's applications, they invariably ask about robots running around our bodies.

My primary goal is to examine the ethical issues arising with nanotechnology: that is, nanoethics. This will not be a listing of what's ethical and what's not, though that will sometimes happen. Instead, I want to explore some of the deeper values and beliefs underlying our fascination with technology and nanotechnology in particular. Before getting into that in depth, two chapters will be spent explaining what nanotechnology is, what is does and what sorts of product are already on the market. This overview will make it apparent that nanotechnology comes in two very different forms that raise different issues for nanoethics. I have chosen to call them 'normal nanotechnology' and 'futuristic nanotechnology'. The distinctions will become clear during the book, but normal nanotechnology covers what most scientists, engineers and businesses are engaging with around the world. This is the normal scientific work bringing new products onto the market every week. This is what many governments are investing heavily in to rebuild their economies.

Futuristic nanotechnology is very different. Although it will be built upon the discoveries of normal nanotechnology, its focus is on literally building new products atom-by-atom using molecular manufacturing. The best-known items in its repertoire are the nanobots and nanites that appear regularly in science fiction. This form of nanotechnology is not on the market, and won't be for a long, long time.

Nanotechnology has become a staple in science fiction. This is what some people think of when they hear about nanotechnology. For nanoethics, this is actually an opportunity. Ethics has been criticized for being overly rational and philosophical. Literature helps bring ethics to life and brings life into ethics. A film or novel engages the heart, mind and soul with the ethical issues. Science fiction is the genre most suited for nanoethics, and Chapter 3 will explain how this works.

The one story most widely associated with nanotechnology is Michael Crichton's *Prey*. It deals with futuristic nanotechnology, and introduces the notion of risk. Chapter 4 will examine some of what we know about nanotechnology risks. We don't have to worry about a *Prey* scenario in the near future, but other challenging issues must be addressed. Because nanotechnology is so new, we know very little about the risks of nanoparticles. Decisions must be made in the face of uncertainty. Chapter 5 examines the precautionary principle, a favoured strategy among regulators for dealing with risk ethically. The debate over its strengths and weaknesses will be examined and some guidelines suggested.

Chapter 6 will put nanotechnology into a global perspective. On the one hand, many governments are looking to nanotechnology to build or rebuild their economies. On the other hand, nanotechnology could be one more way to divide the world into those who have and those who have not. Nanotechnology holds great promise for developing countries, but only if the needs of the poor rank high in how decisions are made about the focus of nanotechnology. Justice is key.

The next two chapters address medical applications of nanotechnology. Already nanomedicine is producing new drugs, diagnostic devices and other methods of benefiting patients. The ethical issues in nanomedicine are currently similar to those in medicine. Challenges will arise as new classes of nano-enhanced drugs enter clinical trials. Moving into Chapter 8 brings enhancement to the fore. This challenging ethical issue is examined in detail because nanomedicine is expected to provide many new drugs and devices that can powerfully enhance the body and mind.

In Chapter 9 we will examine posthumanism, the ideological movement that seeks to enhance humans beyond all current limitations. These views have impacted futuristic nanotechnology, especially in the area of converging technologies: the amalgamation of nanotechnology, biotechnology and genetic engineering, information technology, and cognitive science (NBIC). These powerful technologies have great potential to contribute to human good, but when blinded by the pursuit of perfection, they have the potential to cause much harm. And finally, we round out our discussions by revisiting the Greek myth of Daedalus and Icarus to examine its relevance for today.

Many people have helped me and encouraged me over the years. I will name just a few people here who have had a particular role with this book. The Biomedical Diagnostics Institute at Dublin City University, funded by the Science Foundation Ireland under Grant No. 05/CE3/B754, introduced me to this fascinating topic. Thank you in particular to Brian MacCraith and Kenneth Dawson for encouraging me to grapple with these issues. I am greatly indebted to people who read over drafts of this manuscript. They brought their diverse backgrounds and expertise to bear in helping to improve the manuscript, though of course the faults and views that remain are mine. Thank you sincerely to Helen Berney, Mark Brake, Pat Brereton, Valeria delle Cave,

Richard O'Kennedy, Kermit Horn, Carol Lynam, Sean Ruane, Mike Sleasman and Melissa Stobie. Thank you also to the many others who provided that all-important encouragement to keep going when the eyelids were drooping or the spirit was flagging.

Thank you to Conor and Peter for sitting through all those science fiction films even though you'd rather be watching something else. Thank you to Catrina for your prayers and words of encouragement. And a special thanks to my wife, Cheri. Without your presence at my side this book might never have been written. Without your willingness to give up having me at your side, the book would not have been written. I also want to thank God who gave me the gifts and many opportunities for which I am eternally gratefully. Buíochas mór le Dia.

Dónal P. O'Mathúna
Dunboyne, Co. Meath, Ireland
April 20, 2009

1 Nanotechnology: In Search of a Definition

Everything around us is composed of atoms. We eat them, breathe them and shape them into tools to accomplish our tasks. We are made from atoms. Each new technological age has been characterized by the types of atoms we human beings could manipulate. The next technological age will be characterized, not by the *types* of atoms used, but by the *scale* at which they will be manipulated. This is the essence of nanotechnology. Richard Smalley shared the 1996 Nobel Prize in Chemistry for his work in discovering one of the chemical structures at the centre of the nanotechnology revolution. In 1999, he spoke before a subcommittee hearing of the US Congress:

> From stone, to copper, to bronze, iron, steel, and now silicon, the major technological ages of humankind have been defined by what these atoms can do in huge aggregates, trillions upon trillions of atoms at a time, molded, shaped, and refined as macroscopic objects. Even . . . the smallest feature is a mountain compared to the size of a single atom. The resultant technology of our 20th century is fantastic, but it pales when compared to what will be possible when we learn to build things at the ultimate level of control, one atom at a time.
>
> (Smalley, 1999, quoted in Mansoori, 2005, p. 1)

Definitions of nanotechnology and of nanoscience are not universally agreed upon. The resulting lack of clarity is one factor hindering discussion of nanoethics – the ethical issues with nanotechnology (Schummer, 2007a). Some have argued that the 'nano' label is applied to many conventional areas of research to make projects more attractive to funding agencies. On the commercial end, 'nano' has been added to names of products with the most tentative links to nanotechnology. The Nano became the world's cheapest car when it went on sale in March 2009. Manufactured by Ratan Tata's company in India, its basic model costs what other manufacturers charge to install a DVD player in their cars (O'Connor, 2008). In the native language of Mr Tata's mother, the word 'nano' means 'small' (Chang, 2008).

The term 'nano' comes from the Greek word for dwarf, and in science refers to a part that is one billionth of something (Mansoori, 2005, p. 3). One nanometre (nm) is one billionth of a metre, or 1×10^{-9} metre. Human hairs, with a diameter of about 80,000 nm, or red blood cells, about 7,000 nm in diameter, are much larger than nanoscale objects (Royal Society, 2004, p. 4). Bacteria are larger than nanoscale, falling into the microscale (one millionth of a metre). Individual atoms are smaller, for example, hydrogen atoms are about 0.1 nm wide. If a 1-nanometre particle was magnified to the size of a football, the same magnification would make a football the size of the Earth (*ibid.*).

Chemists have traditionally studied and made molecules which are smaller than nanoscale, molecules like aspirin, penicillin and many other pharmaceuticals. Biochemists study the molecules in living organisms, molecules such as proteins and DNA, life's genetic molecule. Objects that fall into the nanoscale include: strands of DNA (about 2 nm wide); proteins (5 to 50 nm); viruses (about 75 to 100 nm); and a range of new nanomaterials called quantum dots (10 to 100 nm), carbon nanotubes (1.4 nm wide), buckyballs (0.7 nm in diameter) and various nanoparticles such as dendrimers, silica beads, paramagnetic labels and so on (*ibid.*). These will be discussed in the next chapter, as will the development of methods and instruments to study them. Some nanomaterials are already appearing in commercial products, although their potential applications are only just beginning to be realized.

The US National Nanotechnology Initiative (NNI), one of the largest funders of nanotechnology research in the world, uses the following definition:

> Nanotechnology is the understanding and control of matter at dimensions between approximately 1 and 100 nanometers, where unique phenomena enable novel applications. Encompassing nanoscale science, engineering, and technology, nanotechnology involves imaging, measuring, modeling, and manipulating matter at this length scale.
>
> (NNI, 2008b, p. 3)

Not everyone uses the same range, with some definitions including objects with dimensions of 1 to 1000 nm (Ozin *et al.*, 2009, p. 12).

Rather than the precise size of the particles, what is more significant is that particles in the nanoscale range have different properties and functions to those in smaller and larger ranges. Smaller particles are dominated by quantum effects, whereas larger particles are dominated by their bulk properties. The nanoscale range is that in which a unique combination of quantum and macroscale effects converge to give nanoparticles their unique and interesting properties. Another distinctive feature is how nanoparticles interact with cells and living tissues. Small molecules enter cells by passive diffusion depending on their solubility and concentration. Cells grow on or around large structures. In between, nanoparticles interact with similarly sized biomolecules such as proteins to become coated nanoparticles (Lynch and Dawson, 2008). This provides protein-coated nanoparticles with access to active transport systems into cells that open intriguing possibilities for nanoparticles 'permeating the impermeable' (Ozin *et al.*, 2009, p. 571). While this may allow drugs to be directed to exactly where they are needed, it also raises questions about toxicity if nanoparticles go where they shouldn't go.

Sometimes a distinction is made between nanotechnology and nanoscience. For example, the UK Royal Society stated:

> ***Nanoscience*** is the study of phenomena and manipulation of materials at atomic, molecular and macromolecular scales, where properties differ significantly from those at a larger scale. ***Nanotechnologies*** are the design, characterisation, production and application of structures, devices and systems by controlling shape and size at nanometre scale.
>
> (Royal Society, 2004, p. 5)

Part of the complexity of nanotechnology derives from the variety of tools and techniques used and the range of applications involved. Because of this, some use the term 'nanotechnologies' (*ibid.*; ten Have, 2007), but we will use the term 'nanotechnology' as it is more widely recognized.

WHY NANOTECHNOLOGY?

Every material has macroscale or bulk properties we value, ones we can see and experience like strength, weight, colour, conductivity, etc. Steel is strong, but has limitations. It makes cars and aeroplanes stronger, but also heavier and less fuel-efficient. Fibreglass car bodies are beneficial at the fuel station, but detrimental in a head-on collision. Electronic devices from phones to laptops give great portability, but battery life must be balanced against size and weight.

'Materials science' is a field of engineering and applied science which studies the relationship between a material's bulk properties and its underlying atomic and molecular structures. Materials science has led to many developments in metallurgy, ceramics, plastics and semiconductors, and is now playing a central role in nanotechnology.

Nanotechnology could lead to new products with new properties: stronger and lighter materials, faster and smaller electronic devices, cleaner and more efficient manufacturing processes, drugs that go to precisely where they are needed and better diagnostic devices. One website examining nanotechnology's ethics put it this way:

> The significance of this breakthrough, if it becomes a reality, can't be overstated: because everything is made up of atoms, and nanotech plays with those basic building-blocks, it could in theory create or improve anything.
>
> (NanoEthics Group, 2003–2008)

Imagine a material stronger than steel, but as wearable as cotton. Imagine a family car that goes thousands of miles without refuelling or polluting the environment. Imagine a box in every house, similar to a microwave, which could be programmed to make whatever was needed. Fans of *Star Trek* may be reminded of the 'food slots' or 'replicators' which allowed food to be made by arranging atoms and molecules into precise patterns. Similar devices called 'matter compilers' make almost everything needed for the characters in *The Diamond Age*, one of the first novels focused on a society radically changed by nanotechnology (Stephenson, 1995). Such devices are also proposed in serious non-fiction scientific treatises (Drexler, 1986/2006).

Space travel provided the incentive for many recent developments in materials science. One challenge was to build stronger, but lighter space-craft; computers needed to store and process more information, but be smaller in size and faster; space-suits needed to be completely sealed, but more flexible and less cumbersome.

Active imaginations are frequently found around nanotechnology. Claims from both eminent scientists and science fiction authors that we are learning to build things atom-by-atom have led to much speculation about our future with nanotechnology. Will we be able to imagine the properties we want, figure out the nanoscale structures they require, and build the items? Immediately attractive would be drugs to cure all our illnesses, ways to produce all the food and water everyone needs, reliable methods to clean up the environment and economical means of building sturdy housing. Then we could work on cheap computers for everyone, unlimited travel or artificial organs to replace the originals when they fail. Maybe after that will come space travel for the masses, enhancements to improve people's bodies and minds and eventually a cure for ageing and death. If all this is possible, the ethical questions will remain, even as they do in the world of *The Diamond Age*.

> Now nanotechnology had made nearly anything possible, and so the cultural role in deciding what *should* be done with it had become far more important than imaging what *could* be done with it.
>
> (Stephenson, 1995, p. 37)

CHARACTERISTICS OF NANOTECHNOLOGY

Nanotechnology is characterized by the many different fields of study it brings together. Until recently, the natural sciences had been divided up into physics, chemistry, biology, engineering, materials science and information technology, among others. Each focused on different aspects of the natural world, and each tended to function independently and to conduct research in separate university departments. Nanotechnology has developed around questions and issues where these fields overlap. It is one of what are being called Converging Technologies. These are often referred to by the acronym NBIC: nano-bio-info-cogno, or nanotechnology, biotechnology and genetic engineering, information technology and computing, and cognitive science. A report commissioned by the US National Science Foundation concluded that coordinated developments in all these fields 'could achieve an age of innovation and prosperity that would be a turning point in the evolution of human society' (Roco and Bainbridge, 2003, p. x). Such is the scale of expectation.

Nanotechnology means working with substances at close to the atomic scale. The atoms that make up a material are key to its properties. Whether something contains atoms of carbon or atoms of iron makes all the difference. Chemistry studies how atoms combine to form molecules and the bulk properties of those products. Developments in nanotechnology have much to do with how atoms and molecules arrange themselves into structures just above the atomic scale.

For example, diamond is made from carbon atoms. The atoms arrange themselves into a dense structure that gives diamond its very hard and translucent properties. Carbon atoms can also arrange themselves into thin sheets with very different properties. This form of carbon, graphite, is dark and brittle, a useful lubricant that also conducts electricity. In 1985, a completely different form of carbon was discovered called buckyballs. Understanding the structure and properties of buckyballs, and related objects called carbon nanotubes and graphene, is a significant area of research within nanotechnology. We will return to look at these in more detail, but already they are leading to new products and applications.

NANOPRODUCTS

Rarely does the development of a new technology attract the level of interest that nanotechnology has garnered. Some of the interest stems from fascination with humanity's ability to manipulate matter at such a small scale. But much more of the interest stems from very practical developments and expectations.

Nanoparticles are interesting and useful, partly because they have fundamentally different properties compared to larger particles of the same substance. For example, large particles of aluminium are so nonreactive that they are used to make beverage cans we drink from safely. Yet nanoparticles of aluminium are extremely reactive – explosively so. Scientists are interested in them as an alternative fuel since they produce none of the pollutants of fossil fuels (Kleiner, 2005).

Nanotechnology has already led to a host of new products with improved, important and intriguing properties. Some are of more questionable benefit. Nanogum is a chewing gum reported to contain nanoparticles of platinum (katalist, 2008). Regular chewing of this gum is alleged to prevent ageing, a claim which is extremely doubtful (Olshansky *et al.*, 2002a). *Forbes* magazine reported that nanotechnology had allowed confectioners to overcome one of their oldest problems: how to produce chocolate-flavoured chewing gum (Wolfe, 2006). The magazine claimed that nanotechnology overcame the incompatibility of chocolate's cocoa butter with the chewing gum base. The manufacturer later denied that nanotechnology was used in this particular product, but did note that it was using nanotechnology in a number of other confectionery products under development (ElAmin, 2006).

In April 2008, the US-based Project on Emerging Nanotechnologies (PEN) estimated that three or four new nanotechnology products were appearing on the market every week. Over the previous two years, the total number of products had increased from 212 to 609. Between 2007 and 2008, the number of products available doubled, and this rate of increase is expected to continue (Erickson, B. E., 2008). At the beginning of 2009 the total had increased to over 800 consumer products. The one which did most to bring nanotechnology to public attention was Apple's iPod Nano. Its small size and large storage capacity are made possible by the production and precise positioning of electronic components in the nanoscale range.

The amazing developments in electronics are governed by Moore's Law. This isn't really a law, but a trend first observed in 1965 by Gordon Moore, co-founder of Intel. It predicts that the number of transistors that can be placed on an integrated circuit will increased exponentially, doubling every 18 months or so (Roco and Bainbridge, 2003). It has accurately tracked many measures of computing power, and helps drive the search for smaller electronic components. Nanotechnology is expected to push electronics and personal digital devices to their next level of miniaturization. Researchers in Scotland announced that their techniques could increase the storage capacity of an iPod from about 3.3 to 500,000 gigabytes per square inch (Anonymous, 2008). These developments are made possible by breakthrough technology that allows chemical groups to be placed precisely 0.32 nm apart in a nanodevice.

Shortly after the release of the first iPod Nanos, Apple was hit with a class action suit alleging that the products were highly prone to scratching, making their screens illegible (UPI, 2005). Nanotechnology promises to provide better scratch-resistant coatings, with numerous other coatings already using various nanoparticles. Glass impregnated with nanoparticles is marketed for 'self-cleaning' windows. The nanoparticles become energized by ultraviolet rays which break down any organic dirt. The nanoparticles are also water repellent, so that when rain hits the window it spreads across the glass, washing off evenly rather than pooling in droplets and streaking. Elsewhere, silver nanoparticles have been put into bandages and healthcare clothing because of their antimicrobial properties. Nanoparticle coatings have also been put on clothes to make them stain resistant. Researchers have announced a way to use the motion and friction generated by walking to produce small amounts of electricity in people's clothes (Qin *et al.*, 2008). The researchers wove nanoengineered fibres called nanowires around textile fabrics to produce fabric that converts wasted mechanical energy into electricity. With further improvements, they predict that sufficient energy will be produced to enable everyone from hikers to soldiers to power personal electronic devices.

As portable electronic devices become more prevalent, their usefulness can be limited by the weight and longevity of the battery. Nanotechnology is being used to improve the performance of all sizes of rechargeable batteries. One of the first applications of carbon nanotubes has been to improve the longevity of car lead batteries. Conventional lithium rechargeable batteries wear out when put through multiple charging and recharging cycles. Nanoparticles are being used to make batteries that can be recharged many more times (Anonymous, 2006). Car bodies are being reinforced with stronger, lighter nanocomposite materials. Some sunscreens now contain nanoparticles of titanium dioxide to give clear products rather than ones that are white when spread on the skin.

The applications of nanotechnology appear endless. Some of these will be described throughout the book, especially medical applications. Little wonder that nanotechnology has generated great excitement, leading to significant investment by governments, universities and private firms. The US government became the early leader when President Clinton unveiled the National Nanotechnology Initiative (NNI) in 2000 with a $270 million investment (Lane and Kalil, 2005). President Bush continued this investment in 2003 with the 21st Century Nanotechnology Research and Development Act (Allen, 2005). The 2009 budget for nanotechnology reached $1.5 billion, bringing the cumulative investment since the NNI began to almost $10 billion (NNI, 2008b). This represents the largest federally funded, interagency scientific research initiative since the space programme of the 1960s. Many other governments are following suit. In 2004, the total public and private expenditure on nanotechnology was about $3 billion in the US, $3 billion in the EU, $2.3 billion in Japan and $1.9 billion in the rest of the world (Bhushan, 2006).

Nanotechnology has become a focal point of competition between nations and continents. The European Commission (EC) views nanotechnology as providing 'important potential for boosting quality of life and industrial competitiveness in Europe' (European Commission, 2007a, p. 2). Between 2002 and 2006, the EC funding for nanotechnology totalled €1.4 billion, which was one third of the European public funding for nanotechnology; this is expected to almost triple between 2007 and 2013 (*ibid.*). This makes Europe the largest public investor in nanotechnology, although it lags behind the US and Japan in private investment (*ibid.*).

International forecasts suggest the global market for nanotechnology-based products will grow to about $1 trillion in 2015 (National Science Foundation, 2001). This figure is debated, ranging from estimates of $3 trillion to smaller figures quoted by those who claim the larger figure includes products for which nanotechnology makes only peripheral contributions (Nanowerk, 2007). Regardless, while nanotechnology deals with the very small, the stakes are very large.

NANOETHICS

Pursuit of the latest innovative technology is always encouraged with excitement. But innovation includes the perils of unknown risks. As with previous eras of human discovery, the excitement of going where humans have never gone before is taking precedence. But should it? The ethical questions surrounding nanotechnology are several and are only starting to be addressed. Nanoethics is a new field, looking at issues of right and wrong in the development and application of nanotechnology (Lin and Allhoff, 2007).

Nanotechnology is widely acknowledged as raising ethical, legal, social and environmental issues. While much funding is available for scientific research and development, little has been available to examine the health and environmental risks or the ethical concerns. A survey of several journal databases found that, while the number of citations for scientific articles on nanotechnology grew almost exponentially between 1985 and 2001, citations on its social and ethical implications stayed flat and close to zero (Mnyusiwalla *et al.*, 2003). All sides of nanotechnology need to be examined. If not, many of the potential benefits may be lost because of suspicion or fear of what nanotechnology might involve.

Part of the reason why the ethics of nanotechnology have not been scrutinized is that nanotechnology has not captured public attention. Ironically, while governments are striving to make their countries and institutions world leaders in nanotechnology, their citizens, who ultimately fund much of the research through taxes and will be expected to purchase the products, are often unaware of the technologies. Several studies have found that the public is largely unaware of nanotechnology (Currall, 2009). Even with about $50 billion in consumer products on the market, 49 percent of Americans polled in 2008 stated they had never heard of nanotechnology (Hart, 2008). The proportion who had 'heard a lot' about nanotechnology was 7 percent in 2008, 6 percent in 2007 and 10 percent in 2006 (Hart, 2007). Those who know little about nanotechnology tend to view it relatively positively (Scheufele *et al.*, 2007). However, once nanotechnology was explained, consumers tended to become more concerned about its risks than its benefits (Hart, 2008). Some have suggested that the lack of public awareness is good news as it avoids public concern about the new products (Anonymous, 2009). However, fears based on inaccurate information could also damage beneficial developments.

NANOFICTION

This general lack of awareness could be overcome by the use of literature, and particularly the genre of science fiction. Already, nanotechnology has appeared in *Star Trek*, with its on-going conflict with the nanotechnology-enabled Borg, and in other TV series like *X-Files*, *Battlestar Galactica* and *Star Gate*. It has been used in films like *I, Robot* (2004), *DOA: Dead or Alive* (2006) and *Terminator 3: Rise of the Machines* (2003). Probably the best-known work of fiction on the subject has not yet made it to the big screen: Michael Crichton's *Prey* (2002) raises the prospect of nanotechnology run amuck, leading to death and mayhem. The novel has become a metaphor for the potential dangers of nanotechnology, and has led to criticism of how nanotechnology is portrayed in science fiction.

Science fiction often enters discussions of nanotechnology, especially when people fear the vision has run away from reality. If we put new drugs in people, and then little nanodevices, where will it stop? Where *should* it stop? Does the vision for cheap portable electronic devices have anything to do with transporting people around the solar system? Nanotechnology has been plagued with hype regarding its potential benefits and agents of doom suggesting its potential risks (Berube, 2006). Yet science fiction has sometimes described scientific devices that have later been developed, or predicted catastrophes that have materialized (Brake and Hook, 2008). But the flow has not been one way. We will see that nanotechnologists sometimes use fictional accounts to explain their vision – and why they should be funded to try to get there. Some of the devices and scenarios found in science fiction have influenced nanotechnologists and even the design of actual nanotechnology (Berne, 2006).

Given that science fiction engages with nanotechnology, we will explore its potential to encourage careful reflection about nanotechnology and nanoethics in particular. Within the broad field of ethics, narrative and literature are well respected (Booth, 1989; Charon, 2006; Kearney, 2002). Throughout the book, therefore, literature and film will be included in how we examine various ethical issues (Shapshay, 2009a). The specific benefits and limitations of using fiction this way will be considered in a later chapter. Before doing so, we will look at the history of nanotechnology and some of the important subdivisions within the field. Foremost among these are two very different visions for nanotechnology and its potential impact. These two visions must be kept in mind as they raise very different ethical issues.

2 Developing Nanotechnology: In the Beginning . . .

Nanotechnology as a field of research has developed very recently. Much remains unknown about it. The term 'nanotechnology' was coined in 1974 by a Japanese scientist, Norio Taniguchi. He used it to describe the methods and tools used to produce electronic circuits and devices with a precision in the range of 1 nanometre. In spite of the recent arrival of the term, nanotechnology in one sense has existed for centuries. Nanoparticles are produced naturally when materials burn and undergo combustion. They are formed during volcanic eruptions and forest fires, but are also essential to life. As mentioned in the last chapter, substances in our bodies, like proteins and DNA, are nanoscale particles. Chemists have been making nanoparticles for decades in the polymer and plastics industry. The unusual properties of some ancient artefacts are now known to have arisen from the presence of nanoparticles. Growing knowledge of the principles of nanoscience has revealed that down through the centuries, craftsmen stumbled across ways to incorporate nanotechnology into their work without realizing it.

ROMAN NANOTECHNOLOGY

Perhaps the most striking example of nanotechnology in ancient times is the Lycurgus Cup. Made of glass, it depicts mythological characters in exquisite detail (British Museum, n.d.). The method of carving these 'cage cups' dates to the fourth century AD, but few examples survive. They were made by Romans using one of the most technologically sophisticated glass-making processes to be developed before modern times. The most startling feature of the Lycurgus Cup is that it changes colour completely, depending on how light shines on it. When viewed in ordinary light, the cup has a green, jade colour. But when light is put close to the cup and shines out through the glass, the cup has a translucent ruby red colour.

The Lycurgus Cup is the only complete antique item made from glass with a property called dichroism (meaning two-coloured). Since its acquisition by the British Museum in 1958, much work has been done to understand how the colour change occurs (Freestone *et al.*, 2007). The explanation lies in minute amounts of gold and silver found in the glass. A complicated manufacturing process, along with chemical additives, led to the production of distinct silver-gold nanoparticles with dichroic properties. Whether the craftsmen knew how to do this, or just stumbled across this one success, we may never know. A few other fragments of dichroic glass exist, but nothing as spectacular as the Lycurgus Cup.

This antique artefact reveals some of the characteristics of nanotechnology. Properties of a larger item – the bulk properties – sometimes change dramatically because of nanoparticles. In this case, the bulk colour changes because of interactions between light and nanoparticles. These properties are strongly influenced by the size and shape of the nanoparticles. The silver and gold nanoparticles arrange themselves (or self-assemble) into colloids, suspensions of tiny particles which form distinct arrangements or interactions, leading to the bulk properties.

Just as the Roman glass-makers probably produced dichroic glass through trial and error, so some developments in nanotechnology have been serendipitous. Part of the current interest in nanotechnology is driven by the desire to understand how nanoparticles can be designed and manufactured to make products and devices with specified properties and functions.

RICHARD P. FEYNMAN

The vision for 'manipulating and controlling things on a small scale' is often traced to a presentation made by Richard P. Feynman (1918–88) at the 1959 American Physical Society annual meeting (Feynman, 1959/1992). Feynman is one of several scientists associated with nanotechnology who have won a Nobel Prize, though his was for research prior to his engagement with nanotechnology. Feynman has been called 'the founding father of modern nanotechnology', one of the leading physicists of the twentieth century, an outspoken visionary and a true maverick (Gazit, 2007, p. 19).

Feynman's talk did not use the term 'nanotechnology', but inspired research that has contributed greatly to its modern development. However, others who made important contributions to this 'atom-moving business' were not aware of the talk until after they made their discoveries (Berube, 2006, p. 54). Feynman asked why the entire 24 volumes of the *Encyclopaedia Britannica* could not be written on the head of a pin. He calculated that after shrinking the book to fit on a pin head, each letter would still be many atoms wide. That would provide enough resolution to read the letters, provided a tool existed to write that small. Feynman noted that none of this would violate the laws of physics. 'I am telling you what could be done if the laws are what we think; we are not doing it simply because we haven't yet gotten around to it' (Feynman, 1959/1992, p. 61). His confidence in the feasibility of getting huge amounts of information into small spaces, he drew from nature. Throughout biology, cells produce substances, walk around, wiggle, and 'do all kinds of marvelous things – all on a very small scale . . . Consider the possibility that we too can make a thing very small, which does what we want – that we can manufacture an object that maneuvers at that level!' (*ibid.*, p. 62).

We will see that natural, biological objects continue to inspire many of the developments in nanotechnology, as does confidence that humans can replicate, if not exceed, what nature has already accomplished. This confidence is what encourages those involved to work so long and hard overcoming problems with perseverance and creativity. Yet sometimes this can go too far and contribute to ethical problems. The boundary between confident determination and arrogant hubris will be important in our study. Technological accomplishment can lead to technological determinism, a belief that technology shapes cultural values, and that society's problems can be resolved through technological developments (Postman, 1992). This ideology will be especially important in evaluating proposals to use nanotechnology to enhance the human body and take humans beyond their current natural limits (Bostrom, 2008). More on that later, but its origins lie within Feynman's vision.

In his 1959 talk, Feynman went on to describe computers that were much smaller and faster than those available at the time. Current computer technology shows he got that prediction right. He also looked at the possibility of making tiny, moveable machines. Challenges would arise because properties vary in different ways as the dimensions are reduced. Still, he claimed we would figure out ways to 'drill holes, cut things, solder things, stamp things out, mold different shapes all at an infinitesimal level' (Feynman, 1959/1992, p. 63). He believed the same general approach to manufacturing used at macroscale would work at nanoscale. This has become known as the 'top-down' approach to nanotechnology, where bulk quantities of materials are broken down and shaped to give nanoscale pieces.

Feynman also went on to describe what has become known as the 'bottom-up' approach. 'But I am not afraid to consider the final question as to whether, ultimately – in the great future – we can arrange the atoms the way we want; the very *atoms*, all the way down!' (Feynman, 1959/1992, p. 65). Nanoscale devices then would be built atom by atom, molecule by molecule, to give intricately designed objects. He predicted that nanoscale machines could be designed for precise functions. Regardless of whether these machines would be useful or economical, Feynman said they would be fun to make. A friend of Feynman's suggested these machines would have medical uses.

> You put the mechanical surgeon inside the blood vessel and it goes into the heart and 'looks' around. (Of course the information has to be fed out.) It finds out which valve is the faulty one and takes a little knife and slices it out. Other small machines might be permanently incorporated in the body to assist some inadequately-functioning organ.
>
> (Feynman, 1959/1992, p. 64)

While Feynman's vision was percolating through scientific and engineering circles, the idea of swallowing a surgeon was picked up in science fiction. In 1966, the Academy Award-winning film *Fantastic Voyage* was released, along with a novelization of the screenplay by the renowned author, Isaac Asimov. The film contains a number of scientific impossibilities which Asimov (1966) tried to correct, foremost being the actual shrinking of the submarine to get it injected into the patient.

Feynman's approach of devising small machines remains prominent in some visions of nanotechnology's potential contribution to medicine. This approach envisions nanoscale robots (nanobots) that can enter the body and conduct repairs (Freitas, 2005). These tiny machines are described and graphically presented as machines, with gears, pipes, pumps and appendages. They look like robots, beetles or submarines shrunk to size for their fantastic voyages through the body (Viktor, 2005). They also generate controversy, as some nanotechnologists see them as a distraction from the real world of nanotechnology. 'Although these devices are fascinating to consider, they currently represent a form of science fiction rather than emerging reality' (Leary *et al.*, 2006b, p. 1,019).

A major division exists within nanotechnology. On the one hand, developments in nanotechnology are progressing and already producing practical applications, devices and therapies. Workers in this area include many scientists, engineers and doctors who see their research as an extension of other scientific and biomedical research. Their work raises ethical issues similar to those raised by developments in related fields: balancing risks and benefits; environmental ethics; questions of privacy; justice in resource allocation and research ethics in testing new products. We will this call 'normal nanotechnology', in the sense of the relatively routine work of scientists working within established scientific theories (Kuhn, 1962).

On the other hand, some have a much more radical view of nanotechnology that has been distinguished as 'futuristic' (Jömann and Ach, 2006). Claims are made that nanotechnology will allow us to produce anything we want from junk, air and water; eliminate all physical needs through abundance; produce devices and implants that enhance us beyond our wildest dreams; and eventually lead to the end of disease and death. Some even claim we will be able to use nanotechnology to evolve ourselves into a new species, the posthuman (Bostrom, 2008). The ethical issues raised by this approach to nanotechnology are fundamentally different to those of normal nanotechnology, including: the ethics of human enhancement; genetic testing; gene therapy, and attempts to direct our future evolution. This approach we will call 'futuristic nanotechnology'.

These two approaches are different, but interconnected. They each claim nanotechnology will revolutionize the world, but to different extents. They are often motivated by similar visions of technology solving humanity's problems. Both place a high value on pursuing technology, but both differ in how firmly they adhere to technological determinism. We will return to this theme regularly. The two approaches are also related by their reference to science fiction, but the relationship is more complicated than often acknowledged. As exemplified by Feynman, science influences science fiction. The next prominent figure in the history of nanotechnology exemplifies how science fiction influences science.

K. ERIC DREXLER

K. Eric Drexler has become synonymous with futuristic nanotechnology and the idea of molecular manufacturing. During the 1970s, Drexler was an engineering student at Massachusetts Institute of Technology (MIT). He became involved with a number of visionary scientists and activists who were interested in cryonics and space colonization (Regis, 1990). Cryonics involves freezing a person's body (or only the head, if cost is of concern) immediately after death, and thawing it out later when technology has developed the means of correcting whatever caused the person's death. A major problem with cryonics is dealing with the damage caused by the freezing and thawing processes. Drexler began to see how nanobots might solve the problem by infiltrating the body and repairing damage molecule by molecule, cell by cell (Drexler, 1986/2006).

Having seen the potential value of nanobots, Drexler began to envision how they might be built. He looked to nature as a source of inspiration. To show the possibility of building molecular machinery, he made analogies between biochemical components and familiar structures. For example, enzymes have binding sites which hold molecules in place during reactions. Drexler viewed the enzymes as molecular clamps. Other proteins serve to move molecules through membranes, so these he called pumps. By a series of analogies,

> these arguments indicate the feasibility of devices able to move molecular objects, position them with atomic precision, apply forces to them to effect a change, and inspect them to verify that the change has indeed been accomplished.
>
> (Drexler, 1981, p. 5,276)

Molecular complexity in nature was used to validate his proposal that humans could build complex nanoscale machines. For example, ribosomes are where living cells put proteins together. According to Drexler, 'Ribosomes are proof that nanomachines built of protein and RNA can be programmed to build complex molecules' (1986/2006, p. 66).

Biochemistry textbooks use schematic drawings to illustrate how biochemicals work. To help us understand things at a molecular level, these have parts and shapes that are analogous to familiar machines. They are more like literary analogies than real devices. Drexler takes the analogy as reality and proposes to build nanodevices using what he calls molecular assemblers (Drexler, 1986/2006). These instruments would place atoms in precise locations, guided by computer programs carrying the assembly instructions. Once the instructions for any item are figured out, people 'could grow spaceships from soil, air and sunlight' (*ibid.*, p. 163). After refinement, 'assemblers will be able to make almost anything from dirt and sunlight . . . The greatest problem will be deciding what we want' (*ibid.*, pp. 220–1).

Drexler also proposed that nanotechnology pursue 'self-assembly'. He looked to nature again and noted that cells assemble all the components they need themselves. When viruses are taken apart, or protein shapes are distorted, they sometimes self-assemble and restore all their functions. Many structures, from cell membranes to soap bubbles, self-assemble when molecules of the same substances come together in a regular pattern. Drexler predicted that molecular machinery could also be built that would self-assemble, much as bacteria reproduce. Once the process was initiated, the machines would build themselves, thus necessitating little or no human input. This self-replicating capability of nanomachines has become one of the most hotly debated aspects of Drexler's vision. The notion of throwing junk into a molecular assembler and getting useful products out is very different to what happens in cells. To guide biochemical assembly in nature, DNA provides the information and proteins act as templates. Drexler has since acknowledged that self-replicating nanoscale assemblers are not necessary for most of his vision, yet they remain a contentious aspect of futuristic nanotechnology (Edwards, 2006).

Drexler published his ideas in 1986 in a popular book, *Engines of Creation*, reissued in 2006. It has significantly impacted interested scientists, engineers, politicians and science fiction authors, and generated some limited public interest. Drexler also co-founded the Foresight Institute in 1986 to promote public dialogue about nanotechnology and ensure its beneficial implementation.

While Drexler is prominent in promoting nanotechnology, many of those engaged in normal nanotechnology are critical of his ideas (Smalley, 2001). *Engines of Creation* is neither a scientific treatise nor a work of fiction, but it incorporates many literary devices to make rhetorical claims. Some dismiss the book as little more than science fiction, yet that is too simplistic. As we will see later, prominent publications in the scientific and legislative literature have some of the characteristics of science fiction. Perhaps a fairer assessment of Drexler's book is that, while it is not 'science fiction *per se*, it is clearly uncomfortably situated between imagination and reality' (Berube, 2006, p. 55). This has important implications for nanoethics, because some ethical issues are specific to Drexler's futuristic nanotechnology and are not directly relevant to much of normal nanotechnology. Yet ethical concerns about futuristic nanotechnology could negatively impact normal nanotechnology. Nanoethics must be based on an accurate understanding of the type of technology being examined. Therefore, we will return to examine scientific developments to see how far down these two nanotechnology roads we have moved.

PEERING DOWN TO NANOLEVEL

In developing nanotechnology, an important instrument would be one that would allow researchers to 'see' atoms and molecules. In 1981, Gerd Binnig and Heinrich Rohrer, researchers at IBM's Zurich laboratory, invented the scanning tunnelling microscope (Walker, 1990). They astonished colleagues and the public by producing images of individual atoms that were about 5 nm tall. For this, they won the Nobel Prize in Physics in 1986.

The scanning tunnelling microscope (STM) moves an ultrafine tip (or probe) over the surface of substances being imaged. The instrument moves the probe so close to the object that electrons move between them due to quantum effects. This produces a completely different type of microscope compared to optical ones. The STM was the first of a series of closely related microscopes known generally as scanning probe microscopes (SPMs); one particularly powerful instrument now commonly used in nanotechnology is the atomic force microscope (AFM). Each works best in different conditions with different substances, but jointly they allow researchers to visualize the particles and devices they are constructing and studying. As such, scanning probe microscopy has played for nanotechnology a similarly crucial role to that played by light microscopy in the world of cells and micro-organisms.

Not only can an STM image atoms and molecules, but the probe can nudge them around. In 1989, researchers produced an image of the letters IBM written in individual xenon atoms. The feat took Don Eigler and colleagues 22 hours, but they estimate it would now take only 15 minutes (Nanooze, 2005). Within only 30 years, researchers had surpassed Feynman's vision of writing letters that were dozens of atoms wide, as these xenon characters were one atom wide, or about 5 nm. Fascinating and beautiful images generated by STMs moving atoms into specific patterns are now widely available (IBM, 1996).

MAKING NANOPARTICLES

As physicists and engineers were developing ways to picture nanoscale items, chemists were discovering some unique nanoparticles. Richard E. Smalley (1943–2005) has been called the 'grandfather of nanotechnology' (Adams and Baughman, 2005, p. 1,916). In 1985 he was involved in one of the major scientific breakthroughs of the twentieth century, earning him a share of the 1996 Nobel Prize in Chemistry (Ozin *et al.*, 2009). Carbon had been known to occur in three forms: diamond, graphite and powdered carbon (like charcoal), but Smalley, Robert Curl and Sir Harold Kroto discovered a completely different form of carbon. They named it buckminsterfullerene after the architect Buckminster Fuller who designed a spherical geodesic dome for the 1967 World Exposition in Montreal. The structure of buckminsterfullerene is a highly symmetrical sphere of 60 carbon atoms, arranged in repeating hexagons and pentagons. It looks like a football, so the molecules are also called buckyballs.

With a diameter of just under 1 nm, buckyballs have become one of the most thoroughly investigated chemicals of all time and are central to developments in nanotechnology. As a spherical molecular 'cage', buckyballs are being examined for their ability to trap drugs and other chemicals so that they can be transported and delivered to specific parts of the body or locations within cells. Smalley contributed immensely to nanotechnology, being the most frequently cited author in nanotechnology in the ten years prior to his death in 2005. In addition, he was actively involved in political debates on funding nanotechnology. He was one of the few active researchers within normal nanotechnology to publicly engage Drexler in debates over futuristic nanotechnology (Drexler, 2001; Smalley, 2001).

The discovery of buckyballs led to renewed interest in carbon, leading to more developments in 1991. Long cylindrical tubes of carbon were discussed in a paper by Japanese physicist Sumio Iijima in the prestigious journal *Nature* (Iijima and Ichihashi, 1993). These became known as carbon nanotubes, and soon replaced buckyballs as the hot item in nanochemistry. Carbon nanotubes look like a sheet of graphite, rolled up into a long tube. As with buckyballs, the carbon atoms are arranged into hexagons and pentagons. The nanotubes have a diameter of a few nanometres, and can be up to 1 millimetre long.

While Iijima is usually credited with discovering carbon nanotubes, their existence had been noted decades earlier. Two Russian scientists published a paper in 1952 describing hollow nanoscale carbon filaments which were probably carbon nanotubes (Monthioux and Kuznetsov, 2006). At the time, engineers were interested in preventing their formation in coal- and steel-processing plants, so not much attention was paid to them. Their unusual mechanical and electrical properties were not appreciated until recently developed instruments were available to study them.

Carbon nanotubes are predicted to have huge commercial impact in the nanotechnology revolution. They are believed to be the strongest known material on a weight-by-weight basis. They are much stronger than steel, at one quarter its weight. They are also flexible and virtually indestructible (Edwards, 2006). They may play an important role in manufacturing car and spacecraft bodies, and also in lightweight bullet-proof fabrics. They have also been touted as the material by which a space elevator might be constructed.

The notion of a space elevator has been proposed to overcome the cost and danger of getting spacecraft into orbit. The basic idea is that a cable would be extended from Earth's surface to an orbiting space station, and that some sort of cable car or elevator would shuttle people and supplies between the two. The space station would become a base for exploring the cosmos without having to re-enter Earth's atmosphere. Space elevators could thus open up the possibility of everyday space exploration.

The general idea of a space elevator has been linked back to biblical notions of the Tower of Babel or Jacob's ladder to the heavens (*ibid.*). Modern notions are associated with an 1895 proposal by the eminent Russian astronautical theorist Konstantin Tsiolkovsky (Hirschfeld, 2002). Arthur C. Clarke's 1979 novel, *Fountains of Paradise*, began the modern science fiction fascination with space elevators which are now standard items in the genre (for example, Reynolds, 2001). Nanotechnology may bring fiction to reality.

Carbon nanotubes are believed to be the first actual material which could conceivably be strong enough to build a space elevator. A number of private companies are working on various components, although massive obstacles remain to be overcome. One of the prominent companies involved, the LiftPort Group, had scheduled an operational elevator for 2018 (Edwards,

2006). After a crisis in working capital in 2007, the date of deployment has been moved to 2031 (Boyle, 2007). NASA has provided $4 million to a prize fund for competitions to produce the basic building blocks required for a functional space elevator (Spaceward Foundation, 2008).

Carbon nanotubes come in different sizes and shapes which influence their electrical properties. A diameter of a few nanometres allows them to transport electrons and other atoms, leading to suggestions that they may play important roles in energy storage and conversion devices. Some carbon nanotubes act like metals in conducting electricity, but others are semiconductors, having electrical properties in between conductors and insulators. Semiconductors are essential in modern consumer electronics. As small as electronic devices already are, carbon nanotubes could allow even further miniaturisation.

At the moment, however, production of carbon nanotubes is limited by practical difficulties in making pure samples and their cost (Gazit, 2007). Another related material is graphene, a one-atom thick planar sheet of graphite. It is one of the strongest materials in the world and is anticipated to have many electronic applications, earning it the name 'Prince of Electronics' (Ozin *et al.*, 2009, p. 754). Nevertheless, carbon nanotubes will still play important roles in many nanotechnology applications. Given that carbon nanotubes, graphene, buckyballs and diamonds are all made from carbon, nanotechnology is often associated with diamonds. In keeping with eras like the Stone Age and Bronze Age, Ralph Merkle (1997), an associate of Drexler's and prominent advocate for cryonics, has called the coming era of nanotechnology the Diamond Age, given the central role of carbon-based structures in nanotechnology.

With the interest given to carbon nanotubes' great practical potential, other materials were found that rolled up into similar structures. Less well-known metals such as molybdenum, niobium and thallium are commonly incorporated into these structures. These inorganic nanotubes are remarkable lubricants which will likely increase the lifespan and efficiency of motors and engines (Gazit, 2007). Other materials have now been found to self-assemble at the nanoscale into a variety of other shapes and structures. Nanowires, nanoribbons, nanobelts and nanorods are all being investigated for their distinct properties.

In addition to interesting electrical properties, distinctive optical properties were discovered in some nanomaterials, particularly nanoparticles called quantum dots (Edwards, 2006). These particles usually contain a core of semiconductor material like cadmium selenide (CdSe), surrounded by a shell of other material. The resulting nanoparticle is something like a cage (10 to 50 nm in diameter) that traps one or more electrons. One of their most useful properties is that the same material can emit different colours of light depending on its size. The smallest quantum dots emit red fluorescent light, the largest, blue light, and those in between emit all the colours of the rainbow (Ozin *et al.*, 2009). Quantum dots are being developed as nanoscale tags that can be placed on molecules and cells. They hold promise in diagnostic devices, replacing traditional dyes because their fluorescence is very bright, stable and long-lasting. They may also play a role in low-energy light-emitting diodes (LEDs).

Nanomaterials can also have distinctive magnetic properties. Nanoparticles made from various metal oxides (like manganese oxide or iron oxide) can be used in high-density data storage devices and other computer applications (Gazit, 2007). Magnetic nanoparticles are also used as contrast agents in enhanced magnetic resonance imaging (MRI). The nanoparticles have properties that differ from their bulk properties, and these will be examined further in the chapter on their medical applications.

NANOLITHOGRAPHY

So far we have discussed materials that are nanoscale in three dimensions (such as buckyballs or quantum dots), or in two (such as carbon nanotubes). Materials that are nanoscale in only one of their three dimensions include graphene, thin films and engineered surfaces. Advances in this area have been central to reducing the size of electronic devices and computer chips. Layers of substances that are one atom or one molecule thick now are manufactured routinely.

As electronic devices have become smaller and smaller, so methods of fabricating nanoscale circuits and patterns have been developed. Lithography is a general method of using moulds and stamps to produce images and patterns. As devices like the atomic force microscope (AFM) have been developed, patterns and images can now be made at nanoscale. In starting with a block of material and using different instruments to etch out a pattern which can then be used to make a nanoscale stamp, nanolithography is a prime example of the 'top-down' approach to nanotechnology. George Whitesides and his chemistry research group at Harvard University have developed an approach called soft lithography that promises to become an indispensable tool for nanotechnology. Described in 2009 as 'still in its infancy', it provides a rapid, low-cost method of producing high-resolution patterns or complex 3D structures out of a wide variety of materials (Ozin *et al.*, 2009, p. 60). New developments in this area are announced regularly.

THE TWO NANOTECHNOLOGIES

Normal nanotechnology builds upon decades of research and development and has already led to several nanotechnology-related Nobel Prizes. During the 1990s, researchers like Richard Smalley began to address the all-important issue of funding. Governments around the world were petitioned to provide significant investment in nanotechnology. One of the arguments for public funding was that many years of research would be needed to see the practical and commercial benefits of nanotechnology.

This would discourage private companies and venture capitalists from getting involved, as they typically insist on shorter-term returns for their investments.

During the 1990s, the US government invested roughly $200 million per year, while all other governments combined invested about $400 million per year. Each year saw relatively small increases. In 2000, the US National Nanotechnology Initiative (NNI) provided dramatically increased funding, stimulating similar increases around the world. The associated legislation was contained in the 21st Century Nanotechnology Research and Development Act of 2003 (Allen, 2005). This made public the simmering competition between normal nanotechnology and futuristic nanotechnology (Berube, 2006). Drexler and his supporters had argued for inclusion in the legislation of a mandate to test the feasibility of molecular manufacturing. That proposal was removed at the last minute, leaving Drexler's vision on the funding sidelines.

Drexler was understandably annoyed: 'In the U.S., at least, molecular manufacturing became a taboo subject' (2006 edn, pp. 9–10). He held that his vision was used to generate interest in nanotechnology among funders and politicians. Then it became associated, especially in popular culture through Crichton's *Prey* (2002), with 'swarms of ravenous nanobugs' (Drexler, 2006 edn, p. 11). Those promoting normal nanotechnology did not want their projects and products associated with ideas that appeared too risky and futuristic. 'Business and industry fear that claims about nanobots and uncontrolled replication might spur ethicists and environmentalists and cranks to frustrate a powerful economic force in the next few decades' (Berube, 2006, p. 59).

Normal nanotechnology focuses on products that evolve along consistent lines from familiar products already on the market: electronics, special coatings, drugs, and stronger, lighter materials. 'Today, the case for evolutionary rather than revolutionary nanotechnology is the case that may need to be made' (Berube, 2006, p. 64). An American professor of law and technology has noted that 'advanced nanotechnology just seems too, well, spooky' (Reynolds, 2004). At the same time, the subtitle to the 2009 NNI budget is 'Research and Development Leading to a Revolution in Technology and Industry' (NNI, 2008b).

Good, scientific reasons exist to focus on normal nanotechnology. An independent review of the NNI by the US National Research Council (NRC) found it difficult to evaluate the potential of molecular self-assembly in manufacturing. Much of the work to date has been theoretical and based on computer modelling. The potential of futuristic nanotechnology cannot be evaluated because the devices would have to be built 'using tools that do not yet exist' (NRC, 2006, p. 107). The NRC concluded that molecular manufacturing is 'currently outside the mainstream of both conventional science . . . and conventional engineering' and exists in the realm of 'visionary engineering analysis', similar to proposals to build space elevators (*ibid.*).

Apart from the turf-war dimension to this debate within nanotechnology, important ethical issues are at stake. If popular literature raises ethical concerns about nanotechnology, these should not be dismissed by simply labelling them as 'mere' science fiction. Legitimate ethical questions are sometimes expressed through fiction in general, and science fiction in particular. There is also something disingenuous in dismissing the genre when it raises ethical concerns, but then using the language of science fiction when it coincides with a scientific vision.

Science and science fiction have been more interconnected than might be imagined. Science fiction obviously benefits from scientific developments. At the same time, science and technology have benefited from insights first explored in science fiction (Brake and Hook, 2008). The important role that the genre can play in ethical reflection on nanotechnology will be examined in the next chapter.

3 Ethics and Nanotechnology: What's the Story?

Nanotechnology has received great attention in academic circles, and many governments are funding the research. Much hope is being placed in new products and companies that will fuel future economic development. At the same time, the general public appears to have little awareness of nanotechnology. One exception in popular culture is the literary genre science fiction. This may provide a means by which public engagement with nanotechnology and nanoethics can be fostered.

THE LIMITS OF ETHICAL THEORY

Ethics is often seen as a rational approach to moral dilemmas which helps people think through and defend choices about right and wrong. Ethics is sometimes seen as primarily a rational exercise, and often as a means for reason to have victory over the passions. You may feel like stealing, but once you think it through you realize this is not the right thing to do; you may believe it is wrong because it goes against the duties or moral rules you value (reflecting the ethical theory called deontology); or you may take a more utilitarian approach and realize that the consequences of stealing would bring more harm than good. Real life generates complicated scenarios and case studies that challenge our ethics. Careful reasoning helps identify the central ethical principles and precepts which must be prioritized in determining the right course of action.

This view of ethics is important, but has been critiqued as being rather limited. The field of medical ethics has expanded rapidly in recent decades, yet people wonder if it has made medicine any more ethical. Bioethics is a more recent development, addressing ethics in biomedical sciences and healthcare more broadly. However, rather than hearing about biomedical research becoming more ethical, we hear of questionable practices and unethical behaviour. This may reflect increased openness about ethical concerns, not a greater incidence of unethical behaviour. Another suggestion is that ethics has focused too much on reasoning, and that is not enough. People may know the right thing to do, and still choose the wrong thing. Leon Kass, former chairperson of the US President's Council on Bioethics, notes that a purely philosophical approach 'spends little time on what genuinely moves people to act – their motives and passions: that is, loves and hates, hopes and fears, pride and prejudice' (1990, pp. 7–8).

The allusion to Jane Austen's novel is not a coincidence. During Kass's term as chairperson, the Council took the unusual step of publishing a collection of short stories (President's Council on Bioethics, 2003a). This reflects a long-standing tradition that sees an important role for literature in ethics (Booth, 1989; Charon, 2006). The myths of ancient Greece, the parables of Jesus, the fables of Hans Christian Andersen and many other stories testify to the important place of narrative in ethics. More recently, debates over sex and violence on television and in film reflect the realization that stories have a way of teaching ethics even without explicit declarations of what is right or wrong. The structure of a story has a way of communicating ideas, or at least of triggering thoughts and discussions about ethics. As such, fiction can serve an important role in providing 'teacherly texts' (Buckingham, 1987).

The use of novels, films, plays, poetry, songs and other forms of narrative in ethics reflects the belief that ethical decision-making is more than a rational activity. Research in many areas is revealing that 'reasoning comes later and is often guided by the emotions that preceded it' (Brooks, 2009). Ethical decision-making 'is a complicated process of using all of one's faculties – intellect, emotions, and imagination – to evaluate a particular situation and determine how to act' (Hawkins, 2002, p. 76). Historically, ethics has been ambiguous towards, if not suspicious of, emotions, although interest in emotions and ethics is being revived (Nussbaum, 2001). The idea is not new. More than two millennia ago, Aristotle commented that moral education includes learning to 'find enjoyment or pain in the right things; for . . . virtues are concerned with actions and feelings' (1985, 1104b, p. 37). The seventeenth-century French mathematician and philosopher Blaise Pascal stated, 'The heart has its reasons, which reason does not know' (1660, IV: 277). The philosopher Bertrand Russell (1924) stated that 'the heart is as necessary for a good life as the head'.

Narratives help show the emotional, psychological, relational and spiritual dimensions of an ethical situation or question (Robinson, 2006); they can bring to light moral issues previously not seen. In the original *Star Trek* the interplay between Spock, the purely rational Vulcan, and Bones, the emotional doctor, brings to life the debate over reason and emotion in decision-making. Narratives also reveal how a certain attitude, action or word can impact others. They can highlight the ethical issues involved in, and the values intertwined with, technological developments. Fiction thus provides a means of examining our reactions to and beliefs about emerging technologies, with science fiction being particularly applicable to nanoethics.

NARRATIVES AND ETHICS

Narrative ethics is a growing field which covers a diverse range of thinkers and approaches (McCarthy, 2003). As helpful and insightful as it can be, narrative ethics has limitations. It can drift into an approach that amounts to nothing more than expressing one's preferences. Narrative ethics can slide towards relativism and the idea that whatever people choose, that is right for them, so long as it is consistent with their own life stories.

A more moderate and balanced view of 'life as a story' does not have to become arbitrary and relativistic. Our life stories have some of the traits of narrative. They have a beginning, go through various chapters and move towards a conclusion. Ethical issues come to us as part of our narratives. We make sense of them in terms of how they fit or don't fit within our narratives. Sometimes ethical issues arise in isolation, as when we are asked, 'Do you think nanobots should be made and used in medicine?' We may then discuss the pros and cons somewhat in isolation from the rest of our lives.

More usually, ethical dilemmas in real life arise as part of our story, sometimes gradually, sometimes abruptly. We make our way through school and realize others are getting ahead of us using performance-enhancing drugs: What should we do? We get married and have difficulty having a child: Which reproductive technologies should we pursue? Ageing takes its toll on our minds or bodies: How should we respond? Ethical dilemmas arise in the context of our stories, impacted by our values, goals and aspirations, intertwined with other people and influenced by society. All of these dimensions can be found in narratives, though not usually in principles and theories. Literature, then, and fiction in particular, can be more true to life, more rounded, three-dimensional and nuanced than the world of objective, impartial reasoning.

Literature and the arts can engage thoughts, emotions and other senses. All of these serve to stimulate the imagination and lead to a fuller involvement in the story or event (Rethorst, 1997). Before reasoning through the issues, we need to perceive the moral dimension of the situation. Words and images elicit reactions which bring cogency to the events or issues being described. Literature makes the abstract more concrete. In this way, literature brings balance to the cognitive dimension of ethics.

As the imagination is stimulated, readers and viewers are drawn into the events.

> A work of art and particularly one that has a strong narrative quality – whether it be a play, a poem, a novel, a painting, a film, a dance, a piece of sculpture – provides us, then, through its aesthetic power to describe and evoke, a text for both studying and experiencing the moral life.
>
> (Yates, 1988, p. 229)

The images developed in our mind's eye remind us that ethics is not about abstract principles but about real lives and situations. The narrative can pull us into the experience that has led to the ethical question.

A good story can help us identify with a character and see the implications of various beliefs, attitudes or values being lived out, 'we can learn how to live good lives and manage bad situations if we adhere to the lessons and experiences embedded in narratives' (Adams, 2008, p. 179). Wayne Booth (1989) is a literary critic who has long championed the role of narrative in moral formation. He claims that most people can identify some book that has changed their lives. When he identifies with a character in a story,

> I fear that I may be a bit like him and swear that I will work to become less like him. Lord protect me from my own egoistical follies, one may think. In either case, the effect is to reinforce or establish a desire to be the kind of person who does not fall into that kind of trap.
>
> (Booth, 2001, p. 106)

This can provide the motivation that Kass stated is often missing: seeing the right thing to do, we want to do it, and we seek to learn to be more ethical.

Literature shows clearly the relational dimension of ethics. For example, much debate has occurred over the ethics of so-called 'saviour siblings' (Sheldon and Wilkinson, 2004). Parents of a child with a serious illness can use reproductive technology to have another child who will provide bone marrow or other tissue to save the sick child. The ethical debate focuses on conflicting principles and rights, and how best to prioritize them. Jodi Picoult's award-winning novel *My Sister's Keeper* (2004, and film, 2009) explores the lasting personal impact of such decisions on all family members. Without declaring whether the procedure is ethical or not, the novel reveals how it can lead to a complicated network of consequences and ongoing, recurring dilemmas. These must be taken into account in ethics.

Literature reminds us that ethics is embedded in relationships. It undercuts the assumption 'that there is a type of personal morality that only involves the individual's self-interest and well-being without involving others or their well-being' (Yates, 1988, p. 230). Literature shows us that ethics both engages the many dimensions of the individual and includes the many individuals engaged.

At the same time, literature should not be taken to negate the importance of reasoning and reflection on the issues. Having been alerted to the many dimensions of ethics, careful reflection on the rational aspects is still needed. The imagination can help reveal other issues, but resolution still requires the application of general norms and principles, and evaluation of potential consequences.

NANOTECHNOLOGY NARRATIVES

New, emerging and controversial scientific developments regularly provide the plots and narrative devices for science fiction. Nuclear power, cloning, environmental problems and other ethical issues are regularly featured. The genre often raises questions about the inevitability of scientific progress. It reminds us that we are not cogs in a machine but multifaceted beings with cognitive, emotional, relational and spiritual dimensions. Science fiction 'works by conveying the taste, the feel, and the human meaning of the discoveries of science' (Brake and Hook, 2008, p. 216).

Nanotechnology is particularly attractive to science fiction authors. Some nanotechnology will not appear for years, so the scenes and contexts used to examine the resulting devices are far into the future. Graham Collins, an editor with *Scientific American*, notes that 'nanotechnology has become one of the core plot devices on which science-fiction writers draw' (2007, p. 81). The reasons for this are many. 'Sci-fi has always relied on "magic" to a greater or lesser extent, but the possibilities suggested by nano-technology provided writers with a new range of themes and things with which to play' (Erickson, M., 2008, p. 142).

Collins believes that part of the vision for nanotechnology, especially that of futuristic nanotechnology, coincides with a number of central themes in science fiction as a literary genre. The inclusion of nanotechnology gives it the 'appearance of scientific plausibility', something cherished by many authors (Collins, 2007, p. 81). The genre, all the way back to Jules Verne, has been fascinated with machines and nanotechnology provides an endless array. More philosophical issues such as controlling the environment, technological determinism, enhancing human nature and scientific hubris are other common themes.

Henk ten Have (2007), a bioethicist with the United Nations Educational, Scientific and Cultural Organization (UNESCO), has noted that ethical evaluation of new technologies can become a technical exercise of applying moral principles to practical problems in a rational way. The focus can become one of weighing pros and cons, benefits and harms, principles and rights, in an effort to minimize negative consequences. This approach assumes that technologies are value-neutral, neither good nor bad. Ethical evaluation only becomes relevant when the technologies are put into practice, which can be for good or bad ends.

This represents a narrow perspective on technology and ethics. Philosophers have shown that technologies are value-laden and not value-neutral. 'Technology confronts us with moral problems . . . because [technology] penetrates, dominates and even "colonizes" our lives and worlds' (ten Have, 2007, p. 26). Technology is so interwoven into modern life that abstract ethical evaluation of individual technologies is not sufficient. Weighing the pros and cons of a technology must be done, but further ethical analysis is also required.

Every technology is directed towards one or more goals, which confirms that it is not value-neutral (Mitchell *et al.*, 2007). Ethical choices are made when one goal is chosen over another. The goal of developing new drugs, for example, reveals the value we place on good health. It also points to the value we place on achieving good health through pharmaceuticals. Investing in new drugs at the same time means those resources are not available to provide counselling or to donate existing treatments to poorer nations. Each choice promotes certain values, whether intended or not.

Starting from the assumption that technology is value-laden, these values should be made explicit and evaluated. Among the values that can drive technology, ten Have includes a desire for perfection, the promotion of autonomy, differing views of the human body, and the search for immortality. Ethical analysis should also examine these since 'ethics is first and foremost a philosophical endeavour to understand ourselves and our existence in terms of what is desirable or undesirable, supportable or reprehensible, good or bad' (ten Have, 2007, p. 28).

Literature and film provide a valuable way of engaging in this broader type of ethics that ten Have has described. Literature can present the application of technologies in people's lives, as opposed to abstract analyses or context-less cases. Literature can present technology within a more complete portrayal of various consequences than an ethical analysis which simply weighs the pros and cons of introducing new devices and machines. As one science fiction author asks, 'How do we acquire *wisdom* along with all these shiny things?' (Brin, 1990, p. 582). Literature provides one important way.

WHAT SCIENCE FICTION DOES

Not every science fiction novel or film will explore related ethical issues in depth. Some may do so in a superficial or inadequate way. Some films are primarily concerned with being action-packed thrillers or special-effects bonanzas. Some are written primarily for entertainment, and may be viewed as such. Some science fiction explicitly focuses on deeper philosophical and ethical issues. Science fiction is not always the best way to examine nanoethics, but certain stories or scenes provide fruitful avenues by which ethics can be explored. Such exploration may require careful attention to the dialogue provided or be best facilitated by carefully selected discussion questions. Sometimes it just takes thinking about where the author is taking us.

A core feature of narrative is ethical evaluation. 'Storytelling . . . is never neutral. Every narrative bears some evaluative charge regarding the events narrated and the actors featured in the narration' (Kearney, 2002, p. 155). Shakespeare's tragedies, Charles Dickens' novels and *The Matrix* trilogy (1999; 2003) are built around ethical evaluations of people and what they do. 'Far from being ethically neutral, each story seeks to persuade us one way or another about the evaluative character of its actors and their actions' (Kearney, 2002, p. 156).

Character

Science fiction, as a literary genre, can include ethical evaluation of people's character. Such evaluation may have less to do with the futuristic setting and its technology and more to do with the classic themes of fiction. For example, Alastair Reynolds' *Chasm City* (2001) is saturated in nanotechnology: nanoscale implants in everyone's bodies, diamond guns and nano-munitions, buildings that alter and repair themselves and space elevators. Yet the story itself is one of heroes and villains, rich and poor, loyalty and betrayal. Basic human character traits are put into a futuristic context for ethical evaluation. Science fiction writer Orson Card claims his stories were never about futuristic elements. 'Always they were about people and how they created and destroyed each other' (Card, 1990, p. xi).

While *Chasm City* is about human characters and their ethics, the novel can be taken as a critique of technological determinism. The storyline is remarkably similar to an enactment of Drexler's vision of a world with futuristic nanotechnology. Both books describe a future where nanotechnology allows humans to explore the universe. Cryogenics is used to freeze people for their space voyages in search of new resources and places to live. Nanobots pervade the bodies of some, repairing all cellular damage and basically rendering them immortal.

Drexler's vision is one of technological determinism: nanotechnology will solve humanity's problems.

> Accelerating economic growth during recent centuries shows that the rich can get richer while the poor get richer . . . Space resources and replicating assemblers will accelerate this historic trend beyond the dreams of economists, launching the human race into a new world.
>
> (Drexler, 1986/2006, pp. 224–5)

Chasm City shows that even with all the new technology, human greed remains. One man wants his spaceship to take over the fleet. Another seeks revenge on those who killed his family. One colony seeks to take over the others. Drexler places his hope in the rational use of technology to change the human heart. Literature reveals a selfish human heart and asks: Which image rings true? Is it our experience that abundance makes the heart grow more cooperative (as Drexler suggests) or more competitive, as in *Chasm City*?

Dialogue

Ethical reflection may be suggested very explicitly and deliberately, or it may be hidden or even unintended. For example, in

Fantastic Voyage the two doctors in the submarine have different interpretations of the significance of what they see inside the blood stream. Watching the oxygenation of red blood cells, Dr Michaels states that, 'It's just a simple exchange, Mr Grant. Corpuscles releasing CO_2 in return for O_2 coming through on the other side.' In contrast, Dr Duval muses, 'But to actually see one of the miracles of the universe. The engineering of the cycle of the breath.' Dr Michaels replies, 'I wouldn't call it a miracle. Just an interchange of gases. The end product of 500 million years of evolution.' To which Dr Duval comments, 'You can't believe all *that* is accidental. That there isn't a creative intelligence at work.'

The two characters become representatives of two worldviews that often differ in their receptivity to new technology. As such, science fiction can provide settings for discussions that might just as well happen while two characters watch a beautiful sunset or marvel at the graceful motion of a dolphin. Dialogue in science fiction, even if short and relatively superficial, can provide an opportunity to reflect on deeper philosophical issues that intersect with the story being narrated.

Values

Some literature presents ethical issues in more implicit ways. Thus, literature often enacts various ethical positions on an issue rather than declaring which perspective is more ethical. The conclusion is left to the reader or viewer, though often with many hints within the narrative structure suggesting that one perspective is preferable. Michael Crichton's *Prey* (2002) does not explicitly criticize the funding of military research conducted by private corporations, but given the problems this arrangement causes within the novel, the reader is left with many questions about the conflicts of interest inherent in these relationships.

Science fiction is not usually concerned about evaluating specific technologies, which are often more like background props. The focus is usually on the value systems and philosophical beliefs lived out in the stories. This permits a fuller evaluation of their ethical and social implications. Thus science fiction 'stories become modes of power, characters become icons for meaning, and the storyteller approaches the level of poet-priest, speaking Truth through the illusion of fiction' (Collings, 1990, p. 463).

For example, nanotechnology promises new diagnostic devices to monitor the body more regularly and completely. Blood pressure, sleep rhythms, electrolytes, hormones and a host of other markers will soon be measureable in devices as handy and commonplace as a thermometer. Other nanotechnological developments in information technology will allow the collected data to be transmitted wirelessly to a central hub, like a doctor's office.

Such devices will have numerous benefits, such as affording greater privacy, avoiding visits to doctor's offices, and having to wait days for results. But concerns have also been raised, some of which were highlighted in *The Island* (2005). Lincoln Six Echo (Ewan McGregor) awakens to be told he has erratic REM sleep and that he is to report to the Tranquillity Centre. In the bathroom, his urine is screened and he is told he has a sodium excess. Nutrition Control is alerted and he is not allowed to have bacon at breakfast.

This film is dominated by the organization that controls every facet of the lives of those falling under its control. At the same time, Lincoln Six Echo becomes agitated by the constant monitoring, causing him to ask many questions. The director of the Centre listens to his complaints, and then orders more diagnostic tests. Roger Ebert (2005), in reviewing *The Island*, notes that the best science fiction can 'use the future as a way to critique the present'. As such, the film can raise questions about how constant medical monitoring might impact an individual (Kearns *et al.*, 2009). It can stimulate discussion about whether our angst is best responded to with more diagnostic tests or what alternatives should be considered.

Concerns are already being expressed that we live in an increasingly hypochondriacal society. The 'worried well' are constantly checking themselves for illnesses. Might the prevalence of diagnostic devices lead to further anxiety? As we learn more about the impact of stress on health, might not these diagnoses contribute to deterioration in health? Predicting the long-term impact of constant monitoring is difficult, but *The Island* provides one way to look at potential psychological and social impacts, and thereby assists in a thicker ethical analysis.

BACK FROM THE FUTURE

While science fiction often moves into the future, its ethical assessments are frequently directed at the present. Putting current ethical dilemmas into a different context can assist reflection on the issues. Thus, for example, time travel 'has an ideological function because it literally provides the necessary distancing effect that science fiction needs' in order to 'metaphorically address the most pressing issues and themes that concern people in the present' (Redmond, 2004, p. 114). Our current uneasiness and uncertainty about emerging technologies fit well with the concept of travelling to the future. Like the time traveller, we can explore how things might turn out, and return to the present and apply the wisdom we have attained. Ironically, however, while time-travel films have proliferated recently, most do not address the future,

> but deal instead with an escape into an idealised past in a desperate attempt to alter the present and the future. They reflect a growing dissatisfaction with a present that is sensed as dehumanised, diseased, out of control, and perhaps doomed.
>
> (Gordon, 2004, p. 116)

Science fiction looks backwards and forwards to suggest solutions for present-day ethical dilemmas.

A criticism of science fiction is that it can restrict the examination of ethical issues, as some find they cannot relate to the futuristic scenarios. This may be a matter of taste, just as others do not enjoy novels set in Victorian England. By confining the story to a speculative future, the ethical options available to the characters may not work for some audiences if they are too different from today. For example, both *Brave New World* (Huxley, 1932, and film, 1998) and *Gattaca* (1997) explore reproductive technologies, eugenics and technological determinism. The former is set within an authoritarian society where eugenics is state-run and mandatory, making it very unlike current Western society. *Gattaca*, in contrast, deals with many of the same ethical issues, but within a society where the technology is readily available and chosen by individuals. The similarity to modern, liberal society (reinforced by the present-day feel of the film) helps contemporary viewers connect with *Gattaca* more easily than *Brave New World* (Shapshay, 2009b).

SPECULATIVE ETHICS

Nanotechnology gives science fiction authors the opportunity to make predictions about what is to come and to issue warnings. Science fiction is often drawn to dystopian visions of the future, either on an individual level, as in *Frankenstein* (2003), or on a global (or even universal) scale. But it does sometimes paint a utopian vision of humanity transformed into something new and transcendent (Collins, 2007). The *Star Trek* films remain among the few blockbuster science fiction films with a generally positive view of the future (Gordon, 2004).

Alfred Nordmann has raised concerns about the use of speculative arguments in the ethical evaluation of emerging technologies, in particular nanotechnology. He critiques what he calls the 'if-and-then' argument when used to 'invent a mandate for action' (Nordmann, 2007, pp. 32–3). Such speculative ethics can be used both by those promoting and those raising concerns about new technologies. The argument is characterized by a claim about a future development, 'if we could do this', that leads to another claim about what is ethical now, 'then we should do that'. For example, the claim may be made that if we could develop nanotechnology that overcomes ageing, then it is ethical that such research be facilitated now. But someone else might argue that if we could make self-assembling nanobots, they might run amok and destroy the earth, and therefore that it is ethical to restrict their development to prevent this scenario materializing.

Such arguments are commonplace with new technologies. Those promoting their development paint pictures of the future: 'if only' we had a breakthrough. Those concerned about the same technologies paint negative pictures based on 'what if' such-and-such were to occur. Science fiction can then be brought on board to support either utopian or dystopian visions. Others then question whether fictional depictions of the future have anything to offer ethical analysis.

Nordmann makes it clear, however, that the problem is not with using futuristic scenarios, but with the form of the argument and the questions asked. Philosophers have always used hypothetical scenarios and thought experiments to help examine challenging issues. Some stretch the limits of credulity: brains are kept alive in vats to explore their perceptions; a trolley-car rumbles down the track as a passer-by must throw a switch leading to the death of a few or many people; a famous violinist is kidnapped and connected to a patient to serve as a living dialysis machine. While philosophers use these scenarios seriously, 'they do not take them seriously enough to believe them' (Nordmann, 2007, p. 43). Such scenarios are not intended to predict the future, or even describe life in the future. The 'if-and-then' argument, to one degree or another, takes the hypothetical future situation as a given, and from that determines what is ethical today. Such thinking is problematic.

Instead, the purpose of thought experiments should be to raise questions about life as we know it today. Rather than focusing on the possible impact of emerging technologies in the future, we should look at what the speculations tell us about the present, how these speculations critique the present, and why they call for change. Having looked at what the scenarios say about the present, we are in a better position to evaluate the credibility of the claims and their potential to address current ethical dilemmas. 'Science fiction . . . does not prophesy the future; it does, however, lay out possibilities' (Brake and Hook, 2008, p. 253).

Thus, for example, *Spares* (Smith, 1996), *Never Let Me Go* (Ishiguro, 2005) and *The Island* (2005) all examine futuristic solutions to the current problem of the scarcity of organs for transplantation. We may think it unlikely that secret establishments will clone people as spare parts, but the fiction should cause us to reflect on our present situation. Why are there too few organs available for transplantation? In the future settings, cloning is available for the rich and famous. Is that just? Is it any different to how we determine who gets organs today? Given the significance of this problem, is cloning our only option? Rather than shutting down options, science fiction can open up alternative ways to think about situations.

> Indeed, if science fiction scenarios lead to interesting philosophical questions, it is precisely because one suspends disbelief in the presence of fiction. Relieved of the pressure to determine what is true or false, what is likely to happen and what not, we can forge ahead and explore who we are, who we might wish to be, and how these wishes reflect on ourselves or our views of human nature.
>
> (Nordmann, 2007, p. 43)

SCIENCE AND SCIENCE FICTION

Concerns have been raised that ethical discussions of nanotechnology have focused on utopian posthuman descriptions of immortality and bliss or dystopian accounts of nanobots and gray goo. The term 'gray goo' was introduced by K. Eric Drexler and refers to self-assembling nanobots getting out of control and replicating until they devour everything in the world (1986/2006, pp. 355). Crichton's *Prey* (2002) is based around a gray goo scenario involving futuristic nanotechnology.

> These utopian and dystopian visions, which all originate in science fiction, have been reinforced by various others, including futurologists, software engineers, investment consultants, religious sects and governmental agencies. Such visions have created exaggerated public hopes and fears, as well as ethical concerns, with the consequence that nanotechnology is discussed mainly in terms of the societal and ethical implications of these visions.
>
> (Schummer, 2007a, p. 79)

The problem is that attention can be drawn away from the less dramatic and more mundane ethical issues that arise with chemical coatings, nanoparticles and the economic implications of normal nanotechnology. However, science fiction's contribution to nanoethics is not in whether its predictions of gray goo are likely or unlikely, but in examining the values and beliefs upon which nanotechnology can be based. These impact today's normal nanotechnology and thus are ethically relevant.

In addition, the history of nanotechnology reveals a complex and interconnected discourse between science and science fiction. Science fiction does not simply take ideas and gadgets from science and blend them into embellished futuristic stories. In fact, science owes much to science fiction. 'Science fiction helps to train our intellects to accept our imagination as a useful tool within science's toolbox' (Brake and Hook, 2008, p. 253). For example, Feynman wondered about how to build tiny machines that could go into the bloodstream. He suggested making sets of connected 'hands' where the 'slave' is one-quarter the size of the 'master' hands. Each is then wired to more sets of hands one-quarter smaller again. This continues until he has 'a billion little lathes, each 1/4000 of the scale of a regular lathe' (Feynman, 1959/1992, p. 65). As ingenious as the idea may be, science fiction wrote about it earlier. Robert Heinlein (1969) published a short story in 1942 about Waldo, an eccentric mechanical genius who invented hands that could be very large or very small. The smaller ones were used to make even smaller ones until they were small enough to do micro-surgery. They were mounted in banks and operated electrically.

Drexler's self-replicating assemblers have much in common with the nanodevices that get out of control in Crichton's *Prey*. Long before Drexler or Crichton wrote, Stanislaw Lem used self-replicating, tiny machines in his novel, *The Invincible*. Originally published in Polish in 1964, it became available in English in 1973. As with Crichton's nanobots, Lem's act like swarming insects that self-organize and evolve to become destructive to humans.

The relationship between 'normal' science and science fiction is often uneasy. Some regard science fiction as an estranged cousin whom the family has not seen in many years. He seems to have some of the family traits, and knows much about the family history, yet there are other things that raise questions about whether he really belongs or not. Besides that, all he seems to want to do is spin a good yarn, stretching the truth along the way.

Calling something science fiction is sometimes a way of labelling it as unrealistic, detached from reality and not worthy of consideration. Kathryn Cramer has written that, 'The majority of science fiction stories are not plausible extrapolations upon our current situation, using available information' (quoted in Collins, 2007, p. 82). In this approach, science fiction has little to contribute to ethics. Regarding concerns about self-replicating nanobots, Richard Smalley has stated,

> I am convinced that these nanobots are an impossible, childish fantasy . . . We should not let this fuzzy-minded nightmare dream scare us away from nanotechnology. Nanobots are not real. Let's turn on the lights and talk about it. Let's educate ourselves as to how chemistry and biology really work.
>
> (quoted in Schulz, 2000)

Connecting a scientific proposal with science fiction is sometimes viewed as a way to dismiss its credibility. A prominent example is US President Ronald Reagan's vision of building a space-based missile defence system. Technically named the Strategic Defense Initiative (SDI), it was quickly labelled his 'Star Wars' programme. This pointed both to the improbability of the science and engineering and to Reagan's 'Darth Vader' speech two weeks earlier in which he called the Soviet Union an 'evil empire' (Fitzgerald, 2000, p. 22). Nonetheless, the US government has since spent over $60 billion on SDI, making it the single most expensive research project in US history. Critics claim it has yet to produce any realistic benefits (Wright and Gronlund, 2008).

The interconnections between narrative, science and public policy run even deeper. Fitzgerald (2000) reviewed a body of academic work claiming that Reagan's thinking and policies were profoundly influenced by films, especially ones in which he had starred. In *Murder in the Air* (1940), Reagan played an American secret agent with a mission to protect a secret weapon which could destroy all enemy aircraft while they were airborne. An even closer connection can be seen with Alfred Hitchcock's *Torn Curtain* (1966), with its storyline based on the development of an anti-missile missile. Paul Newman played an American agent who states, 'We will produce a defensive weapon that will make all nuclear weapons obsolete, and thereby abolish the terror of nuclear warfare' (Fitzgerald, 2000, p. 23). In Reagan's 1983 speech he called on scientists to develop 'defensive systems' that will make 'these nuclear weapons impotent and obsolete' and thereby 'reduce the danger of nuclear war' (*ibid.*).

Just as Reagan used fiction in developing his science policy, so does nanotechnology sometimes use it to explain or promote its vision. When it predicts advances, 'nanotechnology is thoroughly science-fictional in imagining its own future, and the future of the world, as the product of scientific advances that have not yet occurred' (Milburn, 2002, pp. 266). Even within normal nanotechnology, some descriptions read like science fiction scenarios.

The introduction to one nanochemistry textbook claims that nanotechnology 'is considered poised to revolutionize the world as we know it, and transform us into something better' (Ozin *et al.*, 2009, p. x). Another engineering textbook reads like a scene from *Chasm City* rather than an academic curriculum.

> Computers 1000 times faster and cheaper than current devices; biological nano-robots that fix cancerous cells; towers, bridges, and roads made of unbreakable diamond strands; or buildings that can repair themselves or change shape on command might be future but likely implications of nanotechnology.
>
> (Reith, 2003, p. 2)

Add to these the many claims that nanotechnology is bringing a new renaissance and the next industrial revolution, and Milburn's view can be appreciated better.

Problems arise when science and science fiction are not clearly distinguishable. When scientists use science fiction to describe what is available today, false expectations can be generated. A press release entitled, 'A fantastic voyage brought to life' began:

> Ever since the 1966 Hollywood movie, doctors have imagined a real-life Fantastic Voyage – a medical vehicle shrunk small enough to 'submarine' in and fix faulty cells in the body. Thanks to new research by Tel Aviv University scientists, that reality may be only three years away.
>
> (American Friends of Tel Aviv University, 2009)

It elaborated that 'the real-life medical submarine's Fantastic Voyage won't have enough room for Raquel Welch, but the nano-sized structure will be big enough to deliver the payload: effective drugs to kill cancer cells and eradicate faulty proteins'. For anyone interested in further details, 'The blueprints for the submarine and a map of its proposed maiden voyage were published earlier this year in *Science* by Dr. Dan Peer.'

The actual article in *Science*, however, made no mention of Raquel Welch, submarines or fantastic voyages (Peer *et al.*, 2008).

Instead, a highly technical report described how nanoparticles called liposomes were made with antibodies on their surface that could target specific cells in the body. Inside the liposome was another molecule which could turn off errant RNA in the target cells. In the press release, Dr Peer described these particles as 'biological nano-machines', though liposomes are more like globs of fat trapping other molecules within themselves. The guidance mechanism uses the body's natural immune system, not exactly what the press release conjured up by calling it a 'nano-GPS system'.

As useful as the liposome product may be, its relationship with a submarine stretches the limits of a metaphor. The development is much closer to a group of new drug delivery systems that incorporate nanotechnology. Language should be used carefully and accurately. Metaphors in science 'take what we know and apply it to the unknown and say that the unknown is like the known' (Pitt, 2006, p. 136). But the metaphor goes too far when it suggests that we can do more than we can. When this happens, false expectations develop which benefit no one. We know how to build a submarine, with its metallic shell, engine, communications systems and weapons. We don't know how to build such complicated, intricate devices at the nanoscale. Suggesting otherwise, even in metaphor, is highly problematic.

Science fiction interpreted as a predictor of the future can lead to misunderstandings about nanotechnology. Scientific developments described like science fiction can set up expectations for dramatic breakthroughs. When they do not occur, confidence in the promises of nanotechnology and willingness to continue funding can wane. Even more problematic is the possibility that trust in the scientific enterprise and scientific predictions may be lost. Such trust is going to be very important if the public is to listen to scientists explain the benefits and risks that nanotechnology brings. The question of risk will be examined in the next chapter.

4 Dealing with Risk: Preying on Fear

With any new and emerging field of science or technology, assessment of risk is important, complex and often controversial. Nanotechnology is particularly fraught with such concerns, but with widely divergent assessments. On the one hand, discussions of nanotechnology risks often lead to some mention of Michael Crichton's *Prey* (2002) or some sort of 'gray goo' scenario that will destroy the world. Nanobots, sometimes called nanites, show up more regularly in science fiction, and usually with a negative or dangerous connotation. In *I, Robot* (2004), nanites are used to 'kill' robots; in *The Island* (2005), the nanobots crawl uncomfortably into someone's eye to carry out a brain scan; in *DOA: Dead or Alive* (2006), nanobots are used to secretly record people's activities.

The public perception of nanotechnology could be swayed by these negative portrayals. Concerns have been expressed about potential public reactions to nanotechnology if and when *Prey* is made into a film (*Small Times*, 2004). However, surveys in the US have found that the general public is less concerned than those working in the field about the risks of nanotechnology (Scheufele *et al.*, 2007).

Maybe these dystopian scenarios *are* just science fiction and 'the risks of self-replicating assemblers running amok can largely be regarded as the stuff of fiction' (Gordijn, 2007, p. 111). At the moment, such devices do not exist, but their development is urged by enthusiasts of futuristic nanotechnology. Many of the methods and devices being developed within normal nanotechnology may ultimately contribute to such nanodevices. Collins sees a clear division between the two types of nanotechnology, but that may be overly optimistic.

> The ongoing state of *real* nanotechnology research, for the most part, has little to do with how the details used in fiction evolve over time. The science of science fiction lives in a parallel world to our reality, and what is known to be true or possible in our reality is mirrored only fitfully.
>
> (Collins, 2007, p. 88)

The problem is that history shows otherwise. One of the most notable examples of science fiction predicting scientific risks has been the novel *The World Set Free* by H. G. Wells (1914). There he coined the term 'atomic bomb' for weapons that convert matter into energy via a chain reaction. He predicted that the bomb would be dropped by the Allies on cities in the 1950s with devastating effects. Not only was his fiction realized, but the book inspired Leo Szilárd to mobilize physicists and politicians to develop the atomic bomb through the Manhattan Project (Brake and Hook, 2008). Science fiction sometimes predicts risks very accurately because it understands human nature and natural tendencies.

NANOTECH DYSTOPIA

The general scenario in Crichton's *Prey* (2002) involves a private corporation under contract to carry out research for the US Department of Defense. The research involves genetically modifying bacteria so they produce materials that can self-assemble into nanobots. The public goal is to develop nanobots which will swarm to form a camera for medical imaging, but the hidden agenda is related to military surveillance. The researchers hit a stumbling block and release the nanobots into the environment, hoping they will evolve naturally to overcome the problem.

Instead, they evolve into predatory swarms which become violent and destructive. They take over some humans to avoid being thwarted, and then start killing others. At that point, the thriller takes over as the hero scrambles to save the world.

The novel's concerns about nanobots destroying the world are far-fetched. The plausibility of much of the science portrayed is questionable at best (Phoenix, 2003). At the same time, however, the novel feeds into a general concern about 'gray goo' and the unintended consequences of even laudable actions. Science fiction regularly explores this theme. Although not concerned with nanotechnology, the blockbuster *I am Legend* (2007) starts with humanity on the brink of extinction. The cause is not a freak natural disaster, an alien invasion or power-hungry megalomaniacs, but the most noble of medical research. Having found the cure for cancer, it turns out to have an unexpected side-effect: it kills almost the entire human race. Admittedly an extreme scenario, it reminds us of the uncertainties and risks in developing new technologies. It points to the importance of humility in manipulating nature. Some scientists recognize that if nanotechnology can revolutionize the world, the potential for harm is also great. 'We must make sure this revolution will be used for the benefit of mankind and not for its destruction. As those tools are so powerful, their misuse may actually lead to grave consequences' (Gazit, 2007, p. xiii). As Spider Man's Uncle Ben advised, 'With great power comes great responsibility' (2002).

As discussed in the previous chapter, the value of science fiction lies not so much in whether its predictions are likely or unlikely, but in what the story teaches for today. *I am Legend* reminds us that the good intentions of researchers are not always enough to protect people. Crichton's fiction regularly raises concerns about the unintended consequences of technological advances as well as human greed and hubris (Grimes, 2008). Back in the real world, the pressures on researchers to satisfy funders, meet deadlines and achieve glory have led to violations of ethical and legal codes. Crichton gets that risk right. Regulations and safeguards are needed to help researchers resist temptations from within and pressures from outside to attempt inappropriate procedures.

Researchers are extremely unlikely to do something like release a swarm of experimental nanobots into the environment in the hope that the desired characteristics would evolve by chance (and in time to meet impending deadlines)! The value of Crichton's *Prey* is not in how accurately it predicts the risks of nanotechnology, but in how it alerts people to the importance of taking risk seriously. 'Novels can raise awareness. But only a fool expects a novel to actually answer questions about the wisdom of new technologies' (Reynolds, 2002). Crichton has declared that he is not anti-nanotechnology, but wants to promote dialogue about its appropriate regulation (*ibid.*). As such, even critics of the book's scientific accuracy can agree that it may serve a useful purpose (Phoenix, 2003).

Promoters of futuristic nanotechnology criticize Crichton for creating an overly dystopian picture of nanotechnology, but he is not the only source of catastrophic predictions. In the introduction to the 20th-anniversary edition of his *Engines of Creation*, K. Eric Drexler lamented that popular culture had sensationalized the risks of futuristic nanotechnology. 'Soon, "nanotechnology" was all about making so-called "nanobots" – self-replicating bug-like things that could work miracles, but would inevitably run amok, eat the world, and turn it into "gray goo". And these monster nanobugs were, of course, said to be my idea' (Drexler, 2006, p. 10).

Defenders of Drexler point out that his book described 'gray goo' in only one short passage (1986/2006, pp. 355–6). This part of his vision is said to have been blown out of all proportion in an attempt to characterize futuristic nanotechnology as apocalyptic. Drexler does explicitly describe gray goo in one passage, but his book is littered with references to nanotechnology's potential to annihilate humanity. From the beginning, he claims that nanotechnology will be the greatest technological breakthrough in history, 'for better or for worse' (*ibid.*, p. 58). Later he claims that, 'With assemblers, we will be able to remake our world or destroy it' (*ibid.*, pp. 76–7). He states that what we design with nanotechnology 'may determine whether we survive and thrive, or whether we obliterate ourselves' (*ibid.*, p. 140).

He notes that nanotechnology and advanced artificial intelligence (AI) will give new powers, 'power that can be used to destroy life, or to extend and liberate it' (*ibid.*, pp. 403–4). He claims that advances in software, nanocomputers, replicators, and cell repair machines will bring many benefits. 'Together, these advances will make possible a future rich in possibilities, one of which is our own destruction' (*ibid.*, p. 461). His future with nanotechnology will involve a precarious balancing of risks. 'Replicating assemblers and thinking machines pose basic threats to people and to life on Earth . . . Unless we learn to live with them in safety, our future will likely be both exciting and short' (*ibid.*, p. 353).

Authors of fiction use literary devices to keep readers engaged with their story and eager to turn the next page. Drexler similarly describes the positive dimensions of his vision, but ends chapters with an ominous caveat: 'if we survive' (*ibid.*, pp. 195, 223, 344). He reminds his readers of the 'nauseating possibilities' and 'horrors' of nanotechnology to be examined later in his book (*ibid.*, p. 257). He directs readers to parts of his book where he 'will outline reasons for considering nanotechnology more dangerous than nuclear weapons' (*ibid.*, p. 303). Little wonder that reviewers consider his gray goo scenario more significant than the short section where it is explicitly described. Such is the power of literature to paint a vision using implicit literary devices.

Given that even the principal promoter of molecular manufacturing sees so many dangers with the devices, great caution is needed before deciding to develop them. They open a Pandora's Box of risks that would be difficult, if not impossible, to close. They do not solve a tangible need that cannot be addressed in less risky ways. Their main contribution appears to be in enhancing human capacity beyond what is normal, an ethically questionable goal that will be examined in Chapter 8.

The ETC group (2008), an organization concerned about the global spread of new technologies, has called for a moratorium on research involving molecular self-assembly and self-replication. Science fiction and scientific advocates and critics of molecular manufacturing acknowledge the potential for massive, irreversible damage. Critics from within normal nanotechnology note, 'We are nowhere near making a nanomachine that can assemble arbitrary solids atom by atom, *let alone* pick up a single atom' (Ozin *et al.*, 2009, p. 690). Others claim such nanomachines are technologically unfeasible (Smalley, 2001). The US National Research Council (NRC) concluded that molecular self-assembly lay outside the realm of conventional science and engineering (NRC, 2006). Given the financial constraints faced by the world, and the issues of justice to be discussed in Chapter 6, there seems no good reason to pursue the development of self-replicating nanobots.

RISKS AND CURRENT NANOTECHNOLOGY

While many consider gray goo scenarios as completely unrealistic dystopic science fiction, there is a growing awareness that the risks of nanotechnology must be given serious consideration. The importance of understanding and communicating these risks has been highlighted by regulatory bodies around the world. Unless the risks are accurately understood and carefully evaluated, the public may react negatively to products involving nanotechnology and even to the large public investment in nanotechnology. Researchers, regulators, investors and consumer groups realize that public trust and acceptance of nanotechnology will be crucial for its long-term development and to ensure that its potential benefits outweigh any risks.

A new technology can carry many types of risk beyond just physical risks to human health and the environment. Some broader, social impacts, such as the way enhancement technologies could affect our view of human nature, will be discussed later, but one type of social risk will be mentioned here. As nanotechnology facilitates further miniaturization of digital electronics, they may not be physically risky, but may carry risks for personal privacy (Weckert and Moor, 2007).

Cameras and listening devices that can record people's every movement can be so small that they are almost impossible to see. We do not know the impact these devices will have on individuals or society. *The Final Cut* (2004) explores an aspect of this issue when a memory chip implant can record a person's whole life. Robin Williams plays a 'cutter' who edits the recording after a person dies. Dilemmas arise over what to do when the recording reveals a crime, with Williams earning a reputation for editing skills that make 'saints out of criminals'.

Any technology, such as nanotech chip implants, has the potential for great good or large risks. Constant recording could avert criminal or unethical behaviour because people know nano-CCTV is watching. Yet if this becomes the primary motivation for ethical behaviour, the impetus for moral maturity will be reduced. People could remain at the stage of wanting to avoid being caught, rather than developing the character qualities needed to desire to be ethical.

The complexity of these issues is revealed in *The Island* (2005). After Lincoln Six Echo escapes from the Clinic, his pursuers remember that he is still carrying the nanobots they implanted. As these transmit a wireless signal, they realize they can track Lincoln Six Echo. Already in society we have RFID chips in pets, and they are available for children and adults (McHugh, 2004). How will such devices affect our privacy? How will it impact on people psychologically to know they can be constantly tracked and monitored? Who will be doing this monitoring, and how will it impact society? On the other hand, developments such as social networking and reality TV suggest that people's views on privacy may be changing, and this may have implications for privacy-related ethics.

Science fiction frequently raises questions about privacy and control. *Brave New World* (Huxley, 1932, and film, 1998), *Nineteen*

Eighty-Four (Orwell, 1949), *Logan's Run* (1976), *Total Recall* (1990) and many others explore the ethical principles underlying privacy and can stimulate discussions about our current views of privacy. *Minority Report* (2002) adds another twist to concern about security with a method of monitoring the future to prevent crime. Novels and films like these remind us that any technology, including nanotechnology, can be used for great good or great evil. The same devices that can diagnose or treat illness can be used to violate people's privacy or personality.

These issues point to the complexity surrounding the risks associated with specific types of nanotechnology. In some ways they can be settled by ensuring protection of widely accepted ethical principles and rights, such as that of privacy. On the other hand, they raise broad social, psychological and political questions that go beyond the focus of this book. They point to the importance of a broad dialogue over the type of society we want and whether certain types of nanotechnology will help or hinder us in reaching that vision. Meanwhile, we will focus on the more near-term risks associated with normal nanotechnology.

HEALTH AND ENVIRONMENTAL RISKS

The risks for human health and the environment from nanoparticles currently being produced are of major concern. However, scientific information on these risks is generally lacking (European Commission, 2007b). In some cases, the methods for making these assessments have not been developed or are difficult and expensive. Reliable, readily-available testing methods are needed to determine where nanoparticles are located, their concentrations, and whether they may be causing any harm. Early experiments have sometimes been repeated, producing different results, reminding us of how new nanotechnology is and how much remains unknown.

Cosmetics provide a useful example here. For a number of years, some cosmetic companies have been incorporating nanotechnology in various ways into some of their skin care products. Sunscreens with nanoparticles have been at the forefront of these developments. Sunscreens using traditional bulk UV-blocking materials often leave sunbathers covered in a white layer. When the same chemicals are prepared as nanoparticles, they retain their UV-blocking properties but are transparent when applied. Some claim nanoparticles block UV rays better, but uncertainty remains about their potential risks.

The UK consumer association *Which?* wrote to 67 cosmetics companies asking about nanoparticles in their products (Consumers' Association, 2008). They received replies from 17, of which eight were prepared to provide details. Those companies most commonly added nanoparticles of titanium dioxide or zinc oxide to sunscreens. The organization also examined cosmetics available over the internet and found a wide variety using nanotechnology.

In addition to sunscreens, other products included gold, silver and silica nanoparticles claiming various skin benefits. Two products claimed they contained carbon fullerenes, alleging they had anti-ageing properties. Patents have been issued for cosmetics containing quantum dots and carbon nanotubes, although such products do not appear to be on the market yet (European Commission, 2007b).

The Project on Emerging Nanotechnologies (2009) lists a broader range of cosmetics and toiletries that use nanotechnology. Various water-insoluble vitamins, proteins or lipids are suspended in water-based products using nanoscale emulsions or liposomes (also called nanoemulsions or nanosomes). The nanoscale droplets disintegrate when applied to the skin, leaving no nanoparticles. These include breast creams, make-up, antiwrinkle creams and hair care products. Toothpastes are available that include nanoparticles of gold, silver or calcium phosphate (a natural component of teeth).

From an ethical perspective, two major issues arise. One is the information made available to consumers to help them make informed decisions about products. Members of a Citizens' Panel on nanotechnology drawn from the general UK public 'thought it was unethical not to inform people' of the presence of nanoparticles in consumer products (Consumers' Association, 2008, p. 10). The US consumers' group that publishes *Consumer Reports* contacted five sunscreen producers, and all denied that their products contained nanoparticles. Analyses showed that four of the products *did* contain them (Beveridge and Diamond, 2008). Cosmetics are increasingly marketed on the basis of health claims (that they restore or protect skin from damage or have an anti-ageing effect). Manufacturers should therefore provide information on the nanoparticles contained within their products, especially the evidence to support claims about benefits and potential risks.

The *Which?* report found widely diverging practices with cosmetics (Consumers' Association, 2008). Some companies openly use nanotechnology to market their products, apparently believing that the connection gives their products a competitive advantage. The accuracy of such associations has broad implications. Although not a cosmetic product, a protective sealant called 'Magic Nano' was recalled from the German market after a number of people had breathing problems using it (von Bubnoff, 2006). The ETC group called for a recall of all consumer products containing nanoparticles and a moratorium on all nanotechnology research. Analysis of Magic Nano samples, however, revealed that it contained neither nanoparticles nor any of the other active ingredients listed on its label. This incident led to concerns that legitimate 'nano' products would be harmed by the controversy, but also highlighted the importance of developing and enforcing standards for labelling 'nano' products.

Such scares, coupled with producers not revealing their use of nanoparticles, will create suspicions that could wrongly tarnish the whole industry (Beveridge and Diamond, 2008; Consumers' Association, 2008). Manufacturers differ widely in what they view as acceptable practices. One company reported that they refuse to include fullerenes or carbon nanotubes in their products because information on their safety is incomplete or of some concern, and also stated that they will not use nanoparticles in aerosols for the same reason. Yet other companies are already producing such products. The lack of consistency and, in some cases, openness is clearly problematic.

The second major ethical issue has to do with risk assessment. It is generally accepted that nanoemulsions and nanosomes do not create any new safety issues. They break apart when applied, thereby releasing their ingredients onto the skin. Conventional safety tests should be adequate for these products. For this reason, the European Commission's Scientific Committee on Consumer Products (2007b) stated that the safety of cosmetics containing nanoemulsions and related particles may be assessed by conventional testing methods for bulk products. However, when these formulations are used in medicinal creams, the absorption of the drugs is highly inconsistent, with some formulations enhancing and others slowing down drug absorption.

More concern exists around the products containing nanoparticles which do not dissolve or break apart on the skin. Different particles raise different concerns. The most common nanoparticles in sunscreens are titanium dioxide and zinc oxide. While much is known about their bulk properties, far less is known about their properties as nanoparticles, and some companies did not seem to appreciate the potential differences (Consumers' Association, 2008). While little research is available, experts believe that nanoparticles of these substances are unlikely to raise any new concerns when applied to healthy skin. However, their tiny size may allow them to penetrate damaged skin, including sunburnt skin. Studies have not been conducted here, so the risks are not known. In the USA, the Food and Drug Administration (FDA) does not require any additional testing for nanoparticles of a substance, and this concerns consumer groups (Beveridge and Diamond, 2008). The scientific group advising the European Commission (2007b) has recommended further research into and an assessment of cosmetic nanoparticles on a case-by-case basis.

Higher levels of concern have been expressed about some of the other nanoparticles included in products. Carbon fullerenes have been at the centre of growing unease about their safety. Research is actively on-going, but plausible mechanisms exist by which these particles could be absorbed and cause damage. What might happen after long-term use of these products is not known. Such risks could be acceptable if balanced against significant benefits, but both would need to be defined clearly and accurately. Products with fullerenes are advertised as having anti-ageing effects, yet one expert reported being aware of no evidence to support such claims (Consumers' Association, 2008). In general, experts in the field of anti-ageing note that 'no antiageing remedy on the market today has been proved effective' (Olshansky *et al.*, 2002a, p. 92).

Other nanoparticle products are even more problematic. Silver is known to be antibacterial, and nanosilver is even more potent. Nanosilver is added to toothpaste to kill unwanted bacteria, but our mouths also contain useful bacteria. The consequences of regularly killing both types of bacteria are unknown. In addition, nanosilver is likely to be more readily absorbed through the lining of the mouth, with some swallowed and ingested. The consequences of this for our bodies and natural microbial flora in the gut are unknown. Research is just starting to explore how nanosilver kills bacteria and what risks might be associated with it (Lubick, 2008).

Apart from titanium dioxide, little is known about the toxicity of the insoluble nanoparticles used in cosmetics. Individual products need to be assessed to understand their toxicity. Long-term studies are also needed, because cosmetics tend to be applied repeatedly over long periods of time. No toxicity data is available on the consequences of these products accidentally getting into the eyes, being inhaled or being ingested.

Testing of cosmetics is due to become even more challenging in Europe. Between 2009 and 2013, the European Commission (2007b) will introduce various bans on animal testing of cosmetics. While alternative *in vitro* tests (laboratory tests that don't involve live animals) for bulk products exist, tests for nanoparticles are not yet available. This is particularly problematic, because the toxicity of nanoparticles in animals or humans can differ from what *in vitro* tests predict. For example, a study of three types of nanoparticles tested on rabbit eyes found they had the opposite order of toxicity to what had been predicted based on *in vitro* tests (Prow *et al.*, 2008). Even using known nanoparticles in new organs or tissues can lead to unexpected results. While these factors point to the importance of animal testing, this raises a host of other ethical issues which go beyond the focus of this book.

ASSESSING RISK

Many countries and corporations are investing heavily in nanotechnology and therefore have a large financial stake in ensuring it is successful. An important part of its success depends on ensuring its safety. Unless the risks involved with nanoparticles and nanotechnology are understood and resolved, the technology may never fulfil the promises made in its name or reach its full economic potential. Numerous reports from around the world have urged increased attention to the environmental, health and safety issues associated with nanotechnology. Insurance companies are calling for independent research on the short-term and long-term risks associated with nanotechnology (OECD, 2005). In 2006, UNESCO published a report on the ethics and politics of nanotechnology. It identified both toxicity and exposure to humans and to the environment as the most pressing near-term ethical issues.

Although assessment of risk is an important issue for nanoethics, funding for research on risk assessment and management strategies has made up a relatively small proportion of the nanotechnology-related environmental, health and safety (EHS) budget. For example, the 2009 budget for the US National

Nanotechnology Initiative (NNI) made $1.5 billion available overall. Out of this, EHS research projects are to receive $76 million, or about 5 percent of the total (NNI, 2008b). Other projects might include EHS dimensions, leading to projections that an additional $45 million may be invested in this research. Even so, this represents a small proportion of the investment in what most people acknowledge is a crucial aspect of nanotechnology. Concerns have been raised that even this funding is not focused sufficiently on studies of the effects of nanotechnology on humans and the environment (Hanson, 2008). Much of it is going to develop instruments and tests (NNI, 2008a). This reflects both how young the field is and the need to develop ways to assess the risks. But at the same time, employees are working with nanoparticles and products containing them are on the market.

At this point, much remains unknown about the risks of nanotechnology. Research into these questions is an ethical priority. Some of the issues for which few data are available include:

- How the properties of nanoparticles compare with the properties of the same substances in bulk quantities;
- Whether nanoparticles change in properties when released into the environment;
- The consequences of nanoparticles entering the environment from production facilities, consumer products or disposal of used products;
- Whether nanoparticles combined with other substances are more or less toxic;
- What risks workers face in nanotechnology research and production facilities;
- The health implications of inhaling nanoparticles that become airborne;
- Whether nanoparticles can be absorbed through the skin, and what the resulting effects might be;
- How nanoparticles combine with biological molecules and interact with cells within the body.

Increased awareness of the potential risks is important, but the risks must also be understood accurately. Fear based on unreliable tests or fictionalized futures could prevent the development or acceptance of safe and beneficial normal nanotechnology that reduces current harm. At the same time, minimizing the urgency of risk assessment for normal nanotechnology because science fiction accounts are unrealistic could allow the introduction of harmful products. An accurate and thorough assessment of the actual and potential risks of normal nanotechnology remains in everyone's interests.

EARLY RESULTS FROM RISK STUDIES

Results from studies of nanoparticle toxicity (nanotoxicity) are beginning to be published. These results need to be communicated accurately to the public, an activity well-suited for those engaged in nanotechnology. Surveys in the USA have shown:

> . . . that industry and university scientists are among the handful of groups the public trusts the most for information about nanotechnology – much more than governmental bodies, regulatory agencies and news media . . . Ironically, nanotechnology may also be the first emerging technology for which scientists may have to explain to that public why they should be more rather than less concerned about some potential risks.
>
> (Scheufele *et al.*, 2007, p. 733)

A general concern is that nanoparticles' small size allows them to cross many of the body's protective barriers, leading to toxic effects. Some nanoparticles can penetrate the skin. When ingested, small proportions enter the lymph system, but most are excreted in faeces (raising environmental and public health issues). 'When inhaled, they are efficiently deposited in all regions of the respiratory tract; they evade specific defense mechanisms; and they can translocate out of the respiratory tract via different pathways and mechanisms' (Oberdörster *et al.*, 2005, p. 837). Once in the circulation, nanoparticles are distributed around most tissues and organs. In general, nanoparticles less than 100nm enter cells, those less than 40 nm enter the nucleus and those less than 30 nm cross the blood-brain barrier. The barrier prevents external substances from reaching the brain, and this raises questions about whether some nanoparticles will cross the placenta to reach the developing foetus.

This wide distribution does not mean that nanoparticles in general, or even specific nanomaterials, are toxic. In fact, some manufactured nanomaterials (like carbon black and titanium dioxide) have been used for many years and show low toxicity (SCENIHR, 2009). However, preliminary findings demonstrate that some nanoparticles do have toxic effects. Some toxicity is due to the specific substance, but some effects are due to the size and shape of the nanoparticles. Some early results that reported toxicity have been repeated finding no toxicity, pointing to the complexity of nanotoxicity at present (Barnes *et al.*, 2008). For these reasons, all new nanomaterials should be tested for their toxicity on a case-by-case basis before entering the market or environment (SCENIHR, 2009).

However, pressure exists to continue development and to push ahead with production. Even without the results of safety studies, production is being increased. While nanomaterials are small in size, their general usefulness and wide application are expected to generate a need, and market, for large quantities. For example, about 500 tons of carbon nanotubes were produced globally in 2008, while Japanese companies alone are planning to produce thousands of tons annually within five years (Ray *et al.*, 2009).

> Responsible development of any new materials requires that risks to health and the general environment associated with the development, production, use, and disposal of these materials be addressed. This is necessary to protect workers involved in production and the use of these materials, the public, and the ecosystem.
>
> (*ibid.*, p. 8)

Metallic nanoparticles have, to date, been some of the most frequently used nanomaterials, even though 'little is known about their environmental fate and effects' (*ibid.*, p. 9). Silver, copper, aluminium, nickel, cobalt and titanium dioxide samples were made using both bulk and nanoparticle forms and tested on a number of aquatic organisms (Griffitt *et al.*, 2008). Nanosilver and nanocopper caused toxicity in all organisms at very low concentrations, whereas nanoparticles of titanium dioxide did not cause any toxicity. While the effects on these organisms may not be directly related to how they affect human cells, they do point to their potential environmental impact.

Size changes properties, including toxicity. Bulk gold is chemically inert and very safe. Gold nanoparticles are taken up by cells and accumulate in their nuclei. One study found that spherical nanoparticles were absorbed more efficiently than rod-shaped particles, with cells absorbing 50 nm gold nanospheres more rapidly than smaller or larger spheres in the 10 to 100 nm range (Ray *et al.*, 2009). Although nanospheres were taken up by various human cells, experiments to date have not found any toxic effects. This does not mean they cannot cause damage, but only some preliminary tests suggest that gold nanoparticles may have some detrimental effects.

Gold nanoparticles point to a problem with the current regulations for testing chemicals. In the USA, the Environmental Protection Agency has determined that size will not be a factor in determining what constitutes a 'new chemical', which would trigger more testing and scrutiny (Meyer, 2008). Since many nanoscale materials differ in their properties from bulk materials, concerns have been raised that nanoscale materials will not be tested adequately.

Quantum dots are nanocrystals with unusual quantum effects – they emit prolonged fluorescence of different colours. However, their properties vary considerably depending on the environment in which they are found. This variability makes studying their potential toxicity challenging and few toxicology studies have been published (Ray *et al.*, 2009). Quantum dots sometimes contain highly toxic metals, such as cadmium-selenium cores, surrounded by non-toxic coatings. However, when coated quantum dots were exposed to air or UV radiation for 30 minutes, the coatings broke down, exposing the toxic particles (Derfus *et al.*, 2004).

Carbon nanotubes are emerging as an important new class of materials for their lightness and strength, already used in tennis racquets and bicycle frames. They are long, thin fibres, raising concerns that they might have asbestos-like toxicity and generating fears about their potential impact on workers and the environment (Fernholm, 2008). A study published in 2008 injected particular types of carbon nanotubes into mice who developed inflammation and lesions, as happens with asbestos (Poland *et al.*, 2008). Such damage can lead to mesothelioma, the hallmark cancer caused by asbestos. Damage was found from long, straight nanotubes, but not from the short, tangled nanotubes that are more commonly manufactured (SCENIHR, 2009). The study did not demonstrate that nanotubes cause damage in humans, and injecting them is very different to inhaling them. However, the results lend credence to the possibility that at least some types of carbon nanotubes might be hazardous in similar ways as asbestos.

Another study injected carbon nanotubes into the bloodstream of mice with faulty immune systems (Zhao *et al.*, 2008). After four months, the mice showed no negative effects. However, the carbon nanotubes accumulated in their livers and spleens. This caused no apparent damage, but it is of concern that the particles remained in the mice that long. In contrast, other studies followed radioactively labelled carbon nanotubes and found no accumulation in the liver, spleen and lungs of mice (*ibid.*). These conflicting results point to the importance of developing standardized tests rather than having the confusion of different research groups publishing different conclusions based on different tests.

Carbon fullerenes, or buckyballs, are highly fat-soluble compounds, raising concerns that they might accumulate in fat cells.

This has led to environmental problems in the past when species living in the wild absorbed chemicals from waste disposal or spills. Largemouth bass were exposed to buckyballs in solution for 48 hours. The brains of these fish showed significant oxidative damage. Other organs were not negatively impacted, probably because they have better antioxidant protection systems. The researcher called for similar tests of manufactured nanoparticles claiming that, 'If such preventative principles had been applied to compounds such as DDT and polychlorinated biphenyls [PCBs], significant environmental damage could have been avoided' (Oberdörster, 2004, p. 1,062).

When doctors propose a risky treatment, they can usually tell us we have something like a 1 in 50 chance of this side-effect, and a 1 in 1000 chance of that problem, while on the other hand, if we do nothing we have a 1 in 20 chance of something else developing. Deciding on the basis of such numbers is challenging, but at least the data exist to give us guidance on whether the benefits outweigh the risks. Decisions about nanotechnology are not at this point yet, requiring a different method. One ethical approach commonly proposed is the precautionary principle, which will be the focus of the next chapter.

5 Precaution: More Forwards Slowly

When the projected risks and benefits are uncertain, some recommend 'proceed with caution' rather than 'full steam ahead'. Rather than 'going for broke', many prefer 'better safe than sorry'. As scientific knowledge has increased, our ability to predict the consequences of technological developments appears to have waned. And as the power of technology increases, the risks and hazards of those advances become more serious and global (Beltrán, 2001).

Risk analysis is a well-developed field of study that predicts risks associated with events (Hansson, 2004). For example, the risks of an accident at a chemical factory could be analysed to predict likely consequences. Data on the properties of the chemicals and the structures of the factory would be examined along with the probabilities of various events occurring. Estimates would be made of the likely damage and the costs to humans and the environment.

Predictions about the risks of new technologies such as nanotechnology cannot be made accurately according to the usual methods of risk analysis. With nanotechnology 'so little is known about the possible dangers that no meaningful probability assessments are possible' (*ibid.*, p. 26).

This argument cuts both ways. Just because great good can be envisioned for a particular nanotechnology does not settle the argument that we should therefore push ahead and develop it. In the same way, just because potential harms can be envisioned does not necessarily mean we should not pursue the technology.

And yet many debates proceed as if these approaches did settle the issue. Those promoting a technology will point to its potential benefits and claim the harms are uncertain; those opposing the technology claim the benefits are uncertain and focus on its potential harms. When there are no objective data to examine, the discussion often deteriorates into a power play or a mere appeal to values. How can we deny patients the potential new cure? How can we risk destroying the environment?

Hansson has demonstrated that decisions involving great uncertainty cannot be settled on the basis of risk probabilities. Instead, the structure and validity of the arguments made are more important. Arguments involving uncertainty, such as with nanotechnology, fall prey to what he calls 'mere possibility arguments' (*ibid.*, p. 28). Proponents frequently make two serious mistakes. The first arises from a failure to look at other potential effects of the proposed technology. For example, someone may suggest developing nanodevices that kill cancer cells, and then list their many benefits. This list alone should not persuade us and we should examine their other potential impacts. These should include the harms and benefits, short and long-term impacts, local and global effects, and other potential issues. Once all the relevant effects have been examined, whether to proceed or not will not be settled, but a more informed decision can be made based on a better evaluation of the evidence.

A second mistake that can occur is a failure to consider causes other than the one presented in the argument. For example, someone might state that nanotechnology should not be developed because it will give rise to a 'nano divide' (*ibid.*, p. 33). This refers to a growing disparity between the wealth and opportunities available for those individuals and nations with access to nanotechnology and those without. While such a concern is real, other causes for inequalities should be examined (such as past economic arrangements or patenting regulations). Consideration of these factors might provide ideas on how to introduce nanotechnology without promoting inequalities.

Identifying fallacies in 'mere possibility arguments' will not resolve disputes or dictate the best decision. Instead, it avoids simplistic decision-making based on whether the argument for benefit or for harm is made first or more loudly. In avoiding the above mistakes, more factors will be raised that need to be taken into consideration. Having identified several benefits and harms, steps can be put in place to maximize the former and minimize the latter. In other cases, the harms may appear to be so great, and the benefits so hypothetical, that work on that facet of nanotechnology should be put on hold or abandoned, as was proposed earlier for molecular self-assembly.

HISTORICAL PRECEDENT

Another approach to dealing with uncertainty is to draw analogies to past examples. History reveals several cases where caution appears to have been thrown to the wind, with devastating consequences for people and the environment. Analogies must be used carefully and appropriately. Historical examples, as with fiction, must be applied carefully in ethics and should not be used to predict how nanotechnology will work out, just as science fiction is not about predicting the future. By stimulating our imaginations, history and fiction can reveal values, beliefs, attitudes and general principles that remain relevant or are part of human nature and that therefore need to be taken into account in today's decisions. The more similar the events compared, the more likely the same principles will be involved. However, care must be taken not to stretch the analogy too far and to consider significant differences between situations.

Nanotechnology safety and risk assessment is often compared to that of asbestos, dioxins, Agent Orange or nuclear power (Currall *et al.*, 2006). Without question, the first three have caused serious damage, and this suggests a more precautionary approach should have been applied during their introduction.

Much less was known about the products at the time, pointing to similarities with decisions about nanotechnology today.

We should learn from the past and discern general principles on how best to approach risk when scientific data are incomplete. The European Environment Agency commissioned a study of historical examples where significant harm resulted either when the precautionary principle was not applied (false negatives) or when it was applied unnecessarily (false positives). False negatives result when something is believed to be harmless, but is later discovered to be harmful. False positives result when something believed to be harmful and restricted turns out not to be harmful. The resulting report examined 12 false negatives, but did not find false positives robust enough to warrant inclusion in their report (Harremoës *et al.*, 2001). Future volumes of the report are expected to examine false positives subsequently located.

Asbestos was banned in France in 1997, and the European Union (EU) banned it two years later. Deaths from one form of cancer caused by asbestos, mesothelioma, will continue to rise until 2020–30, probably leading to about 250,000 deaths in Europe alone. Mesothelioma has a latency period of about 40 years after exposure. Producers and regulators of asbestos can be excused for not spotting this risk right away. However, asbestos is harmful in many other ways, especially to the lungs. Early warning signs of the health risks of asbestos were not heeded. 'Looking back in the light of present knowledge, it is impossible not to feel that opportunities for discovery and prevention of asbestos disease were badly missed' – this was stated in 1934 by a former Chief Medical Inspector of Factories in the UK, decades before asbestos use was restricted (Gee and Greenberg, 2001, p. 52).

Asbestos is often presented as raising similar concerns to nanotechnology, especially given the physical similarities between asbestos and some carbon nanotubes and the results of early tests discussed in the previous chapter. Asbestos was first mined in Canada in 1879 and Lucy Deane, a UK factory inspector in 1898, raised concerns about asbestos because of the 'easily demonstrated danger to the health of workers' (ibid., p. 53). In 1906, a report in France described the deaths of 50 women working with asbestos textiles. In 1911, experiments with rats provided 'reasonable grounds' for the harmful effects of asbestos dust. One report after another found evidence of harm, yet production and usage continued to rise, with imports into the EU peaking in the mid-1970s and continuing high until the 1980s (*ibid.*).

While hindsight is helpful, an analysis of this case history demonstrates how the precautionary principle could have reduced the risks involved. In the 1960s, an editorial in the medical journal *The Lancet* argued: 'It would be ludicrous to outlaw this valuable and often irreplaceable material in all circumstances. Situations arise where the use of asbestos can save more lives than it can possibly endanger' (Anonymous, 1967, pp. 1,311–12). Asbestos did prevent deaths through fires and road accidents (as it improved car and truck brakes), and saved energy through improved insulation. It created jobs for workers and generated profits for companies. Just like any new technology or product, a list of potential benefits can be very attractive. The precautionary principle is proposed so that it might prevent the repetition of what has been called a cancer epidemic 'which far exceeds the combined effects of all other known industrial carcinogens' (Peto, 1999, p. 671).

THE PRECAUTIONARY PRINCIPLE

The Precautionary Principle is a general ethical principle used when decisions must be made in the face of uncertainty about the risks of scientific research and development (Myhr and Dalmo, 2007). Basically, caution should be prioritized. The precautionary principle emerged during the 1970s as German forests were seen to be dying (Harremoës *et al.*, 2001). The resulting 1974 German legislation addressed air quality standards before scientific evidence could clearly blame air pollution for the destruction of the trees. Since then, the precautionary principle has been promoted and endorsed as an important guide to ethical decision-making. It was explicitly included in the United Nations' 1992 Rio Declaration on Environment and Development:

> In order to protect the environment, the precautionary approach shall be widely applied to States according to their capabilities. Where there are threats of serious or irreversible damage, lack of full scientific certainty shall not be used as a reason for postponing cost-effective measures to prevent environmental degradation.
>
> (United Nations, 1992)

The Treaty of Maastricht that established the European Community in 1992 stated that the precautionary principle should guide its environmental policies (European Union, 1992). The precautionary principle has been formulated in different ways, and this has led to some criticism of its practicality. Even more problematic, it has sometimes been included without any definition in policy documents.

One influential definition has become known as the Wingspread Statement: 'Where an activity raises threats of harm to the environment or human health, precautionary measures should be taken even if some cause and effect relationships are not fully established scientifically' (Ashford, 1998). Other definitions have been put forward by the World Health Organization (WHO, 2004a) and UNESCO (2005a). The EU broadened its applicability beyond environmental issues.

> Although the precautionary principle is not explicitly mentioned in the Treaty except in the environmental field, its scope is far wider and covers those specific circumstances where scientific evidence is insufficient, inconclusive or uncertain and there are indications through preliminary objective scientific evaluation that there are reasonable grounds for concern that the potentially dangerous effects on the environment, human, animal or plant health may be inconsistent with the chosen level of protection.
>
> (European Commission, 2000, pp. 9–10)

The precautionary principle is one of the underpinning principles of the EU's new regulatory system for chemicals, the Registration, Evaluation, Authorisation and Restriction of Chemicals, REACH (European Group on Ethics, 2007). In 2008, the European Commission adopted a voluntary code of good conduct which explicitly included the precautionary principle to guide research in nanosciences and nanotechnologies (N&N).

> N&N research activities should be conducted in accordance with the precautionary principle, anticipating potential environmental, health and safety impacts of N&N outcomes and taking due precautions, proportional to the level of protection, while encouraging progress for the benefit of society and the environment.
>
> (European Commission, 2008, p. 6)

In contrast, the USA is sometimes said to view the precautionary principle less enthusiastically (Harremoës *et al.*, 2001, p. 12). But while the term itself is not used in US legislation, the general approach has been adopted regularly. Several items were banned in the USA years before Europe adopted similar legislation, including bans on scrapie-infected sheep and goat meat from animal feed, chlorofluorocarbons (CFCs) in aerosols, and the steroid DES as a growth promoter in cows (*ibid.*, p. 12).

The precautionary principle is globally recognized as an important decision-making tool, but its practical implementation remains controversial and hotly debated (Weckert and Moor, 2007). A major problem is the lack of consensus on its definition. Different conclusions and recommendations can arise depending on the particular definition adopted. Some approaches recommend action once risk and safety factors are taken into consideration; others recommend inaction in the face of uncertainty (Phoenix and Treder, 2004). Given its growing use, some of the central criticisms need to be evaluated.

CRITICISMS OF THE PRECAUTIONARY PRINCIPLE

One important criticism of the precautionary principle is its perceived impact on technological progress, that it 'inexorably requires science to be ultra-conservative and irrationally cautious and societies to reject a wide spectrum of possible benefits from scientific advance and technological change' (Harris and Holm, 2002, p. 357). For example, Walter Glannon (2002), a Canadian bioethicist, urges caution in manipulating genes because of uncertainty over the risks. Harris and Holm criticize this approach, calling it 'the paradox of precaution' (2002, p. 356). They accept that genetic manipulation may be risky, but note that failing to act also causes harm: those with genetic diseases will continue to suffer and die. However, nothing in Glannon's approach, or in the precautionary principle, suggests that the risks and benefits of all options should not be considered. Glannon's analysis not only examines the consequences of treating or not treating those with genetic diseases, but also broadens his analysis to the potential consequences for future generations. Having considered all options, he concludes that the risks should lead to a precautionary approach.

Other criticisms arise because of difficulties arriving at a consensus definition of the precautionary principle, which is said to produce impractical vagueness (John, 2007). Harris and Holm (2002) examine the terms in the Rio Declaration where the precautionary principle is invoked for 'threats of serious or irreversible damage'. They note that 'serious' and 'irreversible' do not give clear guidance, which is problematic in public policy. Giving someone a cut that leaves a scar is irreversible damage, but probably not very serious. Something that causes death is serious and irreversible for that person, but they question whether this should always trigger the precautionary principle. If so, they conclude that since some people have choked and died while eating apple pie, the precautionary principle would lead to the banning of apple pie.

The precautionary principle does involve difficult-to-define terms like risk, hazard, cost, benefit, serious or irreversible. This makes all risk assessment challenging. The lack of concise definitions is not necessarily a limitation, and may be an advantage. Blurry boundaries force people to think carefully, and to enter dialogue and discussion which may lead to better understanding. Something like Aristotle's Golden Mean applies here (1,108b–9b). Someone who learns to avoid the extremes of both confidence and fear gains the virtue of courage. Virtue in nanotechnology consists in finding the golden mean that avoids both overly confident decisions to proceed and overly fearful restraint. That does not give a list of boxes to check off, but (as the next section will develop) can promote a virtuous ethos in which research and development serves and protects people and the environment.

Harris and Holm also criticize the precautionary principle for calling for moratoria and prohibitions, 'measures which effectively prevent the possibility of harm' (2002, p. 358). They introduce extreme scenarios and all-or-nothing approaches, such as the apple pie example. In contrast, the EU code of conduct for nanotechnologists states that the precautionary principle involves 'taking due precautions, proportional to the level of protection, while encouraging progress for the benefit of society and the environment' (European Commission, 2008, p. 6). The precautionary principle leads to a broad range of suggested strategies. This acknowledges that every decision has risks and benefits, and that inaction has risks and benefits. A decision not to pursue some type of nanotechnology could lead to harm, either directly or indirectly. For example, one of the reasons given in 1919 to promote asbestos use in public theatres was that if it was not introduced, hundreds of people per year would continue to die in theatre fires (Gee and Greenberg, 2001). Tragically, not taking the precautionary route led to many more deaths.

Arguments against the precautionary principle sometimes involve fallacies. The 'absence of evidence of harm' does not mean that there is 'evidence of an absence of harm'. Yet it is often taken to mean this. The burden is then placed on those who are concerned about harm rather than those who seek to promote a new product. Dietary supplements that include nanoparticles (nanoceuticals) raise particular concerns here. Current US regulations do not require pre-market approval of dietary supplements, as is required with drugs. The burden of proof is then on regulators to show harm to remove products from the market (Erickson, 2009). The precautionary principle would require new products, especially those to be consumed like dietary supplements, to be shown to be safe before entering the market, as occurs with pharmaceuticals (and is still not 100 percent safe given the occasional need for product recalls).

Another issue is the 'healthy survivors fallacy', also called the 'pensioners' party fallacy' (Gee and Greenberg, 2001, p. 60). A false sense of safety can be conveyed when certain people exposed to a risk are presented as healthy and unaffected. The longest-lived person, Jeanne Calment, who died aged 122, smoked for 100 years (Olshansky *et al.*, 2002b). Clearly, this does not refute claims that smoking is harmful, but sometimes this type of fallacy slips in. Retired asbestos workers who continued to show up at pensioners' parties were taken as evidence against the harm of asbestos (Gee and Greenberg, 2001). Those who had been harmed remained invisible for obvious reasons! Assessments of a new technology's potential risks must therefore be conducted properly and carefully, using a variety of approaches, including appropriate statistical analyses.

PRINCIPLES BEHIND THE PRECAUTIONARY PRINCIPLE

Given the importance placed on it by many regulators and legislators, the precautionary principle must be clarified and demonstrated to be practical and justifiable. Researchers and manufacturers do not need abstract or inconclusive rules of thumb, but should be given practical guidance to help with the difficult decisions they must make. Clarity is also needed for consistent and justifiable enforcement, otherwise the regulations and codes run the risk of becoming meaningless rhetoric. While the precautionary principle has been formulated in different ways, several central features can be identified:

- Focusing on situations of scientific uncertainty where important information is unknown.
- Aiming to prevent serious and/or irreversible harm to the environment and health.
- Setting appropriate goals in light of what is known about potential risks and benefits.
- Shifting the burden of proof onto those who seek to develop or market products or processes with significant risks.
- Applying the 'polluter pays' principle when damage does occur.
- Identifying alternative means of achieving goals that carry reduced risk of harm.
- Involving all stake-holders in the decision-making process.

The precautionary principle can be rooted in ethical principles. Foremost amongst these is the notion of 'do no harm', more formally called the principle of nonmaleficence. Developers of new medical treatments must demonstrate that their products are both effective and safe. The precautionary principle applies common sense to new technologies: they should not lead to more harm than good. It provides an important counter-balance to the pressures of bringing new products quickly and profitably to market.

The precautionary principle can be rooted in deontological ethics (John, 2007). This approach points to the duties and responsibilities that arise from respecting persons, nonmaleficence and justice. Other values involved in this approach include recognizing the limits of scientific understanding, and acknowledging the vulnerability of human health and ecosystems.

Utilitarian ethics involves weighing and balancing good and bad consequences. The greatest good for the greatest number of people is a guiding principle in utilitarianism. The precautionary principle can be defended on utilitarian principles, where the risks outweigh the benefits. However, utilitarian approaches run into problems when it is difficult to predict consequences reliably and completely, which is usually when the precautionary principle is invoked. In addition, certain outcomes can be envisioned that are so catastrophic that they should not be counterbalanced by any amount of good outcomes – irreversible damage, for example, such as occurred with mesothelioma and asbestos, or major environmental damage.

Underlying the precautionary principle is the need to accept when we lack sufficient information or insight to proceed safely. 'Precaution gives priority to protecting these vulnerable systems and requires gratitude, empathy, restraint, humility, respect and compassion' (Schettler and Raffensperger, 2004, p. 66). Although these virtues are widely respected, they can go against the forces pushing for innovation and novelty. Yet with new technologies, 'their very novelty might be taken as a warning sign' (Harremoës *et al.*, 2001, p. 170). At the same time, similarities with existing products with more data should be taken into account. The toxicity of asbestos has been linked to its long, thin fibres that can be inhaled and are durable. Investigations into carbon nanotubes are needed to determine if they are more or less like asbestos. The last chapter mentioned some early results suggesting they may have some similar properties which should encourage greater caution in their production and use. The long latency period with asbestos mesothelioma points to the importance of not waiting until harmful effects register in population studies before taking precautionary steps. Calls for certainty about risks before exercising caution are unreasonable in the midst of uncertainty. As we realize what we don't know, we should be more humble in our predictions and how we proceed.

Uncertainty should lead to an appreciation of other voices.

Early warnings about asbestos were not heeded, in part because they were raised by women and workers with little power (Gee and Greenberg, 2001). Nanotechnology has developed on the basis of collaboration between scientific and engineering disciplines. Other voices concerning the environmental, social, ethical and legal aspects of nanotechnology should be sought out and listened to. Fresh eyes can bring new insights to a subject. People from outside nanotechnology may also have insights into how the developments might impact on the real world.

The scenarios in science fiction may not only be speculative and imaginative, but also realistic and valuable in suggesting risks. The novel *Chasm City* is set in an environment very similar to that predicted by Drexler (Reynolds, 2001). Nanodevices run throughout human bodies, yet no one seems to have anticipated an alien culture carrying a viral infection that causes the nanodevices to run amok. While extended space exploration is futuristic and the alien culture far-fetched, our current experience with computer viruses should alert us to the potential risks of nanodevice viruses. The lack of attention in the fictional world to an obvious risk should leave us more diligent in looking for the obvious that we may have overlooked in our world.

The precautionary principle does not try to stop or slow science, but encourages it in certain directions. To better understand the risks and benefits, more research is needed on nanoparticles. With new products, research is needed before they are rolled out, and detailed monitoring afterwards. In historical cases when harm has occurred, a general assumption existed that 'if there were harmful effects, evidence would emerge of its own accord and in good time for corrective action' (Harremoës *et al.*, 2001, p. 172). This is not always the case. The precautionary principle therefore strongly supports and encourages research like nanotoxicology.

As research on risks proceeds, early warnings should be taken seriously, especially if based on reasonable mechanisms of action. Long-term adverse effects will be harder to determine, but short-term damage must be investigated. This did not happen with asbestos, and there were serious long-term consequences. If evidence of harm becomes available, final proof should not be necessary before restricting products. On the basis that the scientific case for antimicrobial resistance was not strong enough, concerns about widespread use of antibiotics in animals and humans did not lead to precautionary steps for years (Edqvist and Pedersen, 2001). The precautionary principle eventually prevailed and led to reduced antibiotic usage in animals, and now the causal connection is firmly established.

Research is usually confined to the laboratory, but devices *are* put to use in the real world. New technologies need to be tested in these conditions. New underground storage containers for petrol were introduced to reduce risks from leakage, but were later found to cause more damage because they were not installed properly (von Kraus and Harremoës *et al.*, 2001). PCBs (polychlorinated biphenyls) cause serious environmental damage, but continue to be allowed in 'closed' systems, apparently without recognizing that these get old and leak (Koppe and Keys, 2001). No regulations can control against theft, smuggling or counterfeiting, yet these are real-world risks that need to be incorporated into precautionary thinking.

Precautionary approaches include examining alternatives and their respective potential impacts. Asbestos remained on the market, in part because its cost was kept artificially lower by not including the cost of environmental and health dangers. This made it harder for better substitutes to get on the market. Imagination may be needed to think about alternatives, especially those that are not technological. As widespread use of antimicrobials in animal feed was reduced, alternative farming practices were sought. Swedish farmers brought their real-world farming knowledge to bear in reducing large-scale use of antimicrobials (Edqvist and Pedersen, 2001).

Many forces are at play as nanotechnology develops. Short-term business and political interests may be in conflict with long-term societal and environmental interests. Regulatory agencies must therefore maintain independence from economic and political special interests. In many of the case studies examined by the European Environment Agency, interested parties influenced regulators unduly. Decisions were not based on available evidence, sometimes because of 'persistent obstruction and misinformation by vested interests' (Harremoës *et al.*, 2001, p. 179). Therefore, the precautionary principle incorporates the 'polluter pays' principle. If damage *does* occur, whether anticipated or not, the costs should be borne by the producers. While this has a punitive dimension, the hope is that it will lead developers to assess risks more carefully before releasing products.

CONCLUSION

Recognition of human nature should lead to humility. Just as we do not have certainty in our knowledge, so we cannot guarantee that people will do the most ethical thing. The global economic crisis that began in 2008 is a reminder that sometimes people do what will bring themselves power and wealth, even if it puts many others at risk. Regulations help to reduce some of this, as can the precautionary principle. It is not perfect and it does not dictate what must happen with every technology, but it is a useful decision-making strategy with a strong ethical basis that can help reduce some of the risks of nanotechnology.

All forms of technology require balancing risks and benefits. However, modern technology has introduced products and processes that carry risks of unprecedented proportions that require fresh ethical analysis (O'Mathúna, 2007c). Biotechnology, and nanotechnology in particular, 'raises moral questions that are not simply difficult in the familiar sense but are of an *altogether different kind*' (Habermas, 2003, p. 14). Not only do our technologies manipulate nature, some want nanotechnology to manipulate human nature. This raises new kinds of risk we will consider in later chapters.

Risk assessment must incorporate not only the costs and benefits, but also the place of responsibility (Jonas, 1984) and the building of trust between society, industry, academia and government (O'Neill, 2002). That trust has been shaken by a number of accidents, tragedies and atrocities caused by chemicals and technology. Some of these disasters can be traced to a bias inherent in science and apparently accepted by society. New technologies always seem to be safe when there has been little time for harm to reveal itself. We seem to tolerate 'false negatives' much better than 'false positives' which generates more risks (Harremoës *et al.*, 2001, p. 184). The precautionary principle could go a long way to overcoming this bias. Then 'society would gain overall from a more ethically acceptable and economically efficient balance between generating false positives and false negatives' (Gee and Greenberg, 2001, p. 60).

Precautionary principle assessments of the potential toxic effects of nanotechnology on humans and the environment are challenging. The assessments are further complicated by broader psychosocial concerns. One of these is the impact of nanotechnology on the global economy and the economies of individual countries. This is usually presented as a major benefit, yet nanotechnology is not without its economic risks, especially for developing countries, as the next chapter will show.

6 Global Nanotech: Turning the World Upside Down

Previous chapters have noted the claims that nanotechnology can have a global impact. While much of the research is currently happening in the developed world, the results could affect developing countries. K. Eric Drexler's vision for nanotechnology is of an era of abundance where everyone will be healthy and wealthy. His futuristic nanotechnology involving molecular assemblers 'will be able to make almost anything from dirt and sunlight' (Drexler, 1986/2006, p. 220). Thus, he claims, the poverty and disease that plague developing countries will be wiped away.

Those developing normal nanotechnology also see great potential in assisting developing countries. Nanotechnology could provide ways to supply clean water; improve food supplies; generate cheaper, cleaner energy; make portable information and communication technologies available inexpensively; and prevent, diagnose and treat the diseases of poverty (Peterson and Heller, 2007).

The potential impact of nanotechnology on developing countries raises a number of ethical issues. The suitability and meaning of the term 'developing countries' are much debated. Many rich countries are developing, depending on how the term is defined; and some poor countries are not developing in certain ways or can be losing ground. The term will be used here to reflect countries that are in a low to medium state of development according to the most widely accepted index used by the United Nations. This index takes into account per-capita gross domestic product (GDP), life expectancy and educational standards. 'According to that index, the least developed countries are all in sub-Saharan Africa to be followed by South Asia, Arab states, East Asia, and Latin America' (Schummer, 2007b, p. 292).

NANOTECHNOLOGY IN DEVELOPING COUNTRIES

The US National Nanotechnology Initiative (NNI) has declared that nanotechnology has the potential 'to bring about the next industrial revolution' (NNI, 2003, p. 1). However, revolutions often have casualties. The economic impact of nanotechnology could bring great benefits and new wealth to poorer countries, but as industry changes, economies grow and decline. Developed countries are experiencing the pain of industrial changes as jobs move to other parts of the world. The relationship between economic developments and down-turns is highly complex. Nanotechnology might bring about much industrial change, but that does not determine who will benefit. One theory of economics called dependency theory raises serious concerns about market forces. 'If nanotechnologies have a potential for an [sic] legitimate industrial revolution . . . the dependency theory would predict that, all else [being] equal, they would reinforce the divide between the rich and the poor' (Schummer, 2007b, p. 295).

Some developing countries are following the lead of US, EU and Japanese governments by investing heavily in nanotechnology. China, South Korea and India have large nanotechnology programmes already producing commercial products; Thailand, the Philippines, South Africa, Brazil and Chile have made large investments; Argentina and Mexico are getting established (Court *et al.*, 2007). Governments, universities and industries are coming together in many initiatives. These decisions reflect a recent trend in developing countries to invest in science and technology as a way to spur economic growth. While these trends provide a welcome move away from a North–South divide (the northern hemisphere having greater science capacities than the southern), they raise concerns about a growing 'South–South gap', with Sub-Saharan Africa and Arab countries being on the losing end (Hassan, 2005, p. 65).

Markets change as technology develops, and not everyone benefits. Many African countries produce tantalum, used in capacitors for electronics – but nanotechnology seems better suited to ceramic capacitors. Chile has benefited significantly as a producer of lithium for rechargeable batteries – but nanotechnology-enabled hydrogen fuel cells could replace those. The lighting industry has relied heavily on tungsten, 90 percent of which is mined in China. Light-emitting diodes (LEDs) and other nanostructured filaments will greatly impact the market for tungsten.

As states consider investing in nanotechnology, developing countries cannot afford to make mistakes. They have less income to use in the investments, and can ill afford to go down dead-ends or to attempt projects that do not realize practical benefits. Investment decisions should be based on resources available locally and should address local needs. For example, many developing countries have mineral resources that are or will be of greater value as nanotechnology develops. Pursing developments based on these resources could benefit local economies more than nanotechnology requiring expensive imports.

LEARNING FROM THE PAST

Industrial developments have often not benefited developing countries as much as they could.

> In the twentieth century, many developing countries exported cheap ores, unrefined metals, crude oil, and so on, and imported expensive refined metals, alloys, petroleum, plastics, and so on, leading to increasing trade deficits and astronomic debts.
>
> (Schummer, 2007b, p. 301)

Careful attention to ethical principles will be needed to ensure that change occurs in ways that actually benefit developing countries. Nanotechnology could repeat past mistakes, but it has the potential to promote new standards for ethical business practice.

Other problems have occurred in previous interactions between developed and developing economies. Research and development inherently involves the unknown. Nanotechnology will involve uncertain risks. Local or international investors may be tempted to set up production plants in countries with less restrictive regulations or ineffective regulatory enforcement. As more becomes known about the risks of nanoparticles and how to protect against them, that information must be taken into account in developing countries. At the same time, overly protectionist, paternalistic approaches should be avoided. As discussed in the previous chapter, the precautionary principle can be used to help find a golden mean that provides benefits ethically.

Weak regulatory infrastructure or workers' willingness to accept riskier jobs may make some countries attractive for nanotechnology developments. This may remove incentives to enhance regulatory oversight or promote safer work environments. The primary concern should be worker and environmental safety, with similar regulations adopted and promoted internationally. This requires an approach to involvement in developing countries that accepts an ethical responsibility to promote local development, not exploit local resources purely for financial gain.

NEED AND RESEARCH

The needs of developing countries are significant. Over 2.5 billion people (almost 40 percent of all human beings) live in severe poverty; about 18 million of these people die prematurely every year from poverty-related causes (Pogge, 2008). At least 1.1 billion people lack access to safe drinking water, the main cause of diarrhoea in developing countries, leading to about 2 million deaths per year (WHO, 2004b). Every 80 minutes, the same number of people who died in the Twin Towers on 9/11 die around the world because they happened to be born where they don't have access to food, water or basic healthcare.

At this point in the development of nanotechnology, decisions are being made about the types of projects to fund. Difficult decisions must be made about what is feasible and what is needed. Many issues go into these decisions. Executives want to see products come to market and bring a return on investment. Governments want to see new businesses established and jobs created. Academics want to produce new knowledge and publications. Justice and care of the poor should not be ignored.

Early developments in nanotechnology do not reflect a concern for the poor. 'As is the case with other technologies, advances in nanotechnology tend to be geared to the interests of industrialized countries' (Court *et al.*, 2007, p. 156). In 2005, *Forbes* magazine published what it regarded as the top-ten new nanotech products of the year: the iPod Nano, cooking oil, chewing gum, anti-ageing face cream, baseball bats, strain-resistant fabric, odourless socks, stain-resistant paint, self-cleaning glass and an air purifier (Wolfe, 2006). In 2009, *Forbes* listed five breakthrough technologies predicted to have a major impact in the next decade. Most are nanotechnology-dependant: building materials with built-in solar panels, personal genome sequencing, quantum dot medical imaging, graphene memory chips and multi-touch displays (Wolfe, 2009). The extent to which any of these will help the poor is questionable, raising serious concerns about the criteria upon which their impact is rated. Is it new knowledge, innovativeness or profit? Justice would insist that their impact be measured against how they improve the human condition, especially for those with the greatest need.

The challenges in developing countries are large and complex. For example, some nanotechnology research has addressed the need for safe water. Membranes and filters made from carbon nanotubes, nanoporous ceramics and other nanomaterials are being tested (Meridian Institute, 2005). Some filters remove not only all contaminants, but also all essential minerals, leading to nutrient deficiencies. So far, these devices have been too expensive for many developing countries, and have been hindered by a lack of information on their safety (Hillie and Hlophe, 2007). Issues with technology transfer and local acceptance have also created obstacles, leading Schummer to conclude that 'nanotechnology-based water purification has largely failed' (2007b, p. 297).

Technology alone will not answer humanity's most pressing problems. Nanotechnology's interdisciplinary emphasis could be a major advantage here. People from various disciplines need to address the complex problems in developing countries. However, if nanotechnology over-emphasises its claim to revolutionize the world, it may lead to the neglect of basic, everyday ways to address multi-faceted needs. For example, low-tech, inexpensive water purification methods such as solar disinfection (leaving water to sit in the sun in plastic bottles for several hours) or chlorination have been found to increase water quality effectively and reduce the incidence of diarrhoea (Clasen *et al.*, 2007). The search for new technology must be balanced with investing in efforts to improve existing strategies and to understand the important underlying socioeconomic issues.

HEALTH NEEDS AND NANOMEDICINE

As will be examined in the next chapter, nanotechnology is expected to develop many new medical and pharmaceutical products. Billions are being invested in nanomedicine, already leading to annual sales in billions of US dollars (Resnik and Tinkle, 2007a). For the reasons that will be developed in this chapter, nanomedicine should be targeting the so-called 'diseases of poverty' (Stevens, 2004).

A rough estimate of current interest in a health-related research topic can be gauged from a search of PubMed.gov.

This electronic database holds 18 million citations to research articles on medicine, healthcare and biotechnology. Researchers in relevant fields want to ensure their articles are listed in PubMed to give them greater exposure. In April 2009, a search of PubMed using the phrase 'developing countries' found large interest: almost 74,000 articles. Using only the term 'nanotechnology' returned 15,639 articles. Searching for articles on 'nanotechnology' and 'developing countries' found nine articles. Using the term 'developing world' with nanotechnology revealed five articles, all but one of which had already been identified. Thus, ten articles focused on developing countries, or 0.06 percent of all nanotechnology articles.

Admittedly, this is a very crude estimate of interest in a research topic. Many articles present only the scientific data without reference to where the results might be applied. Research carried out in or for developing countries may not have included the above search terms. However, even if the estimate is off by ten-fold or even a hundred-fold, it still points to a lack of attention within nanotechnology for the research needs of developing countries. This contrasts with the hope expressed by nanotechnology researchers that, 'The resulting medical breakthroughs will benefit the global village' (Sanhai *et al.*, 2008, p. 244). Looking back at how medical research has been targeted in the past raises serious questions about whether developing countries will be included within the boundaries of this village.

RESEARCH DISCREPANCIES

The health needs of developing countries are broadly recognized. What should raise ethical concerns is the disparity between health risks in different countries. For example, life expectancy in some developed countries is more than double that in developing countries: Japan, 82.3 years; Hong Kong, 81.9 years; Iceland, 81.5 years. This contrasts with Zambia, 40.5 years; Sierra Leone,

40.7 years; Zimbabwe, 40.9 years (Rennie and Mupenda, 2008). Child mortality is more than 90 times higher in some developing countries than in developed countries, and the chances of a woman dying during pregnancy are 1 in 7 in Malawi compared to 1 in 2,800 in developed countries (*ibid.*).

Eighteen million people die prematurely each year – roughly one third of all deaths – from medical conditions for which cures exist (WHO, 2004b). About 11 million of these deaths are of infants and children (Pang *et al.*, 2004). In 2002, it was estimated that about 4 million people died from respiratory diseases such as pneumonia, 2.8 million from HIV/AIDS, 2.5 million from pregnancy-related complications, almost 2 million from diarrhoea, and over 1 million from malaria (WHO, 2004b). Although meeting healthcare needs is more complicated than simply delivering treatments, many of these deaths could have been avoided if existing, effective treatments had been provided.

Other diseases exist for which adequate treatments are not available. Diseases like dengue fever, river blindness, sleeping sickness, Chagas disease, elephantiasis and schistosomiasis infect millions of people annually, claiming many lives and inflicting untold pain and suffering (Tropical Disease Research, 2005). For some, treatments are available but have serious limitations or side-effects. For others, no treatment or diagnostic test is available. These infectious diseases primarily affect poor people in the developing world and urgently need the attention of researchers. They have been labelled 'neglected diseases' because despite 'an ever-increasing need for safe, effective, and affordable medicines for the treatment of these diseases, drug development has virtually stopped' (Trouiller *et al.*, 2002, p. 2,188).

Huge investments continue to be made in medical research. Between 1975 and 2004, a total of 1,556 new pharmaceuticals were marketed (Chirac and Torreele, 2006). One percent were directed at neglected diseases, despite the fact that these ailments constituted 11.4 percent of the total global disease burden. All 16 new drugs for neglected diseases developed before 1999 were later listed on the WHO Essential Drugs List, an indication of the significance of their impact on people's lives (Trouiller *et al.*, 2002). Less than 2 percent of all the other new drugs developed at that time made it onto the WHO list. In addition, two thirds of the new drugs were later evaluated as providing little or no therapeutic gain over the products already on the market. The motivation for their development appears to have been more about generating new patentable products than relieving disease and suffering.

The discrepancy between disease burden and investment in research has been called the 10/90 gap. The term was coined to convey the findings of a report published in 1990 by the Commission on Health Research for Development. This landmark report found that while 93 percent of the burden of premature mortality is borne by the developing world, only about 5 percent of the world's investment in health research is directed towards the health problems of the developing world. The term has been expressed in a number of different ways: that only 10 percent of the world's investment in health research is directed towards 90 percent of the world's health problems or disease burden (Ramsay, 2001); or that 10 percent of global health research funding is spent on diseases that afflict 90 percent of the world's population (Vidyasagar, 2006).

The 10/90 gap seeks to draw attention to global inequities in investment in health research. The continuation of this injustice leads to harm across the world. For example, while malaria, pneumonia, diarrhoea, and tuberculosis are among the leading causes of avoidable death, and together account for 21 percent of the global disease burden, yet they receive only 0.3 percent of all public and private funds invested in health research (Global Forum for Health Research, 2004). Because of these gross disparities, there is an urgent need to examine ways of bringing about change so that the lives of those being harmed by the current situation can be improved.

MORAL MOTIVATION

Given the devastation caused by disease, some are saying that the current system of health research funding and reward leads to the violation of people's human rights. Thomas Pogge is a political scientist and a vocal critic of the current system. He has stated, 'The governments and citizens of the high-income countries could and should know that most of the current premature mortality and morbidity is avoidable through feasible and modest reforms' (Pogge, 2005a, p. 199). As such, the current situation is an example of global injustice.

The fundamental reason why the developed world should promote research, including nanotechnology, on conditions affecting the developing world is a moral one. People with few resources need the help we can give. Pogge puts it very bluntly:

> If citizens in the affluent countries were minimally decent and humane, they would respond to these appeals and would do their bit to eradicate world poverty . . . and, seeing how cheaply this can be done, we surely have positive duties to do so.
>
> (2005b, pp. 35–6)

Pogge (2008) calculates that shifting 1 percent of aggregate global income from affluent nations to a special fund would be sufficient to eradicate world poverty. This would include $20 billion to incentivize health research on neglected diseases. Most people in developed countries could afford to do without that 1 percent. The year 2000 marked the first time in human history that the number of extremely poor people (1.1 billion) was matched by 1.1 billion overweight people (Worldwatch Institute, 2000). Both groups suffer from malnutrition, yet at the same time much food is wasted. In the USA, up to half the food harvested never gets eaten; in the UK, about one third of all food is never eaten, wasting about £20 billion each year (Heap, 2005). That is wrong.

As countries are finding billions to bail out ailing banking systems, they still cannot come up with the millions they had promised to developing countries. The UN secretary general Ban Ki-moon has stated that the current financial turmoil could be 'the final blow that many of the poorest of the world's poor simply cannot survive' (Fitzgerald, 2009). Those approaching ethics from vastly different perspectives come to the same conclusion. In biblical language, 'if you know the good you ought to do and don't do it, you sin' (James 4:17). The same view is expressed by Peter Singer, an ethicist and vocal critic of Christian ethics: 'if it is in our power to prevent something bad from happening, without thereby sacrificing anything of comparable moral importance, we ought, morally, to do it' (1972, p. 231).

In an era when ethics is often boiled down to cost-benefit analyses, many are struggling to find reasons to care for the vulnerable. When people's value is determined by their contribution, or decisions are based purely on economics, our sense of justice for the poor is squelched. Jürgen Habermas, the German atheist philosopher, says that justice is rooted deeply in Judeo-Christian ethics. Though society claims to have moved on from religious ethics, Habermas concludes that notions like human rights and democracy are 'the direct heir to the Judaic ethic of justice and the Christian ethic of love . . . To this day, there is no alternative to it' (Habermas, 2006, p. 150). We have the sense of what is right, but have rejected the source of moral motivation to do what is right.

WHERE HAVE THE BIOETHICISTS GONE?

Even the writings of bioethicists appear to have a 10/90 gap. Although little empirical study has been conducted on the issues bioethics addresses, a review of nine leading bioethics and ethics journals found that 96 percent of the articles came from authors in developed countries (Borry *et al.*, 2005). This has led to suggestions of a 'first-world bias' in bioethics (Rennie and Mupenda, 2008, p. 2). Bioethics tends to address topics relevant to affluent countries with access to high-technology medicine and biotechnology. Issues that impact the poor in developed or developing countries are addressed rarely. If even bioethics is not examining the concerns of the poor, little wonder that scientific research is not focused on that area. The reasons for this are complicated.

> Perhaps one reason bioethicists are reluctant to address global ethical issues related to health, illness, and poverty is that bioethicists are deeply embedded in a global economic system that depends on the continued existence of impoverished societies. While there are many ways for corporations to generate profits, one effective means is to shift factories and jobs to places where employers are relatively free of government regulations and where labourers work for a pittance.
>
> (Turner, 2004, p. 175)

Some of this could be down to a lack of awareness. Many wealthy people do not go where the poor live. Film and fiction can help here to create awareness of ethical issues. Documentaries and films can trigger emotional reactions that overcome rational neglect. *The Constant Gardener* (2005) brought to public attention concerns with the way clinical trials are conducted in developing countries. A film like *Slumdog Millionaire* (2008) can do much to put the plight of the world's poor on people's minds and in their hearts. But once the emotional shock or empathy subsides, further reflection is needed to understand the ethical problems and to find informed, appropriate ways to get involved in making a difference.

TAKING STEPS

Nanotechnology cannot solve all the problems of global poverty. Even if nanomedicine provides new ways to prevent, diagnose and treat 'diseases of poverty', many other challenges remain. Some hold that 'many of these [nanotechnology] developments seem so high-tech that it is hard to imagine their being used as health interventions among the poor' (Meridian Institute, 2005, p. 7). Affordable products are needed, which must then be made available to those in need.

One third of the world's population lacks access to the essential drugs that already exist (Pang *et al.*, 2004). Eleven million children die annually in the developing world, two thirds of whom could be saved by available, effective, low-cost interventions such as vaccines, vitamins and insecticide-treated bed-nets (*ibid.*). Once drugs are available, systems must be established to ensure they can be acquired, stored and distributed properly (Bell, 2005). Problems occur in maintaining supply quality. Counterfeit drugs are widely available in developing countries, causing widespread harm and loss of confidence in modern healthcare products (O'Mathúna and McAuley, 2006). Regulatory systems that effectively guard against such deceptive products need to be developed and maintained.

Broader social factors complicate things further. Dealing with disease also requires examining housing, hygiene, sanitation, food storage, transport, communications and many other factors. Here again, nanotechnology could be used to develop products that overcome current deficiencies. Nanotechnology-enhanced sports equipment has been to the fore of new products; some of these developments could provide more durable and lighter equipment for agriculture or fishing. The question is whether such applications will be considered, or whether the focus will remain exclusively on products that make profits. Some have argued that nanotechnology will fail to achieve its potential unless issues of global social justice are not pursued with equal vigour to the scientific challenges (Sreenivasan and Benatar, 2006).

Many of the challenges are not with technology itself (O'Mathúna, 2007b). Technology and information must be made available in culturally appropriate ways. Patenting procedures should allow new products to be available in timely and affordable ways in developing countries, and not create barriers, though this in itself is a large and complex topic (Pogge, 2008; Fender,

2007). Training and education must take account of local situations. For research and development to address local needs, collaboration must be encouraged with local researchers: problems have occurred in medical trials ranging from misunderstandings to clinical trials that viewed participants as guinea pigs for studies that would not have received ethical approval in developed countries (O'Mathúna, 2007a). This has led to a lack of trust in some communities that will take time to rebuild.

A number of strategies have been developed and implemented successfully already, especially with drugs for HIV/AIDS. Nanotechnology programmes could learn from these in many ways. For example, educational institutions could pursue collaborative arrangements with universities and research institutions to facilitate research development. Early career researchers could benefit from mentoring, with a goal of helping them develop local research infrastructure. Care needs to be taken to ensure that such programmes do not lead to a nanotechnology 'brain-drain capacity-building conundrum' (Rennie and Mupenda, 2008, p. 6). As developing countries have invested in training healthcare professionals, many have been lured away to practise in developed countries. While the issues are complicated, vacant positions in developed countries must be balanced against the acute need for well-trained personnel in developing countries, especially if the local community has invested scarce resources in their training and education. Research is needed to demonstrate which strategies are most effective in developing the desired infrastructure, yet that has rarely been conducted (Hyder *et al.*, 2003).

All of these issues are complicated and difficult. Some question whether it is worth trying to tackle them at all, especially 'in light of the incompetence, corruption, and oppression prevalent in so many poor countries' (Pogge, 2005b, p. 47). The suffering and death of men, women and children are enormous. Nanotechnology is being hailed as ushering in a new renaissance, a new industrial revolution (Roco and Bainbridge, 2003; NNI, 2003). If it is to elicit any change in the world, it must focus on the deprivation that already exists. The needs of people in developing countries should rank high in setting priorities for nanotechnology research agendas, and more broadly in the global research agenda.

We might not be able to eliminate poverty and disease, but we should see it within our moral duty to take what steps we can towards that goal.

> Yes, some will get away with murder or with enriching themselves by starving the poor. But this sad fact neither permits us to join their ranks, nor forbids us to reduce such crimes as far as we can.
>
> (Pogge, 2005b, p. 53)

The question for nanotechnology is whether those in decision-making roles will find the moral courage to adapt or adjust their projects to help the developing world share in its fruits, or whether nanotechnology will become one more measure of a widening gap between those who have and those who have not.

7 Nanomedicine: Honey, I Shrunk the Doctor

Medical improvements are an area where nanotechnology promises much and is already beginning to deliver (Sanhai *et al.*, 2008). As of 2007, about 200 companies were involved in nanomedicine research and development, 38 products were on the market with dozens more in the pipeline, and annual sales had reached $6.8 billion and were expected to double by 2012 (Resnik and Tinkle, 2007a). Even with these developments, nanomedicine is a small area within nanotechnology. Of all nanotechnology publications, about 4 percent focus on nanomedicine (Wagner *et al.*, 2006). While the USA and Europe each contribute about one third of these publications, the USA contributes over half of the patent filings for nanomedicine, with Europe generating about one quarter.

Nanomedicine, like nanotechnology, has been difficult to define concisely. Cells and tissues contain nanoscale structures such as proteins and DNA. Since conventional medicine sometimes deals with these materials, clear boundaries between medicine and nanomedicine are difficult to draw. However, nanotechnology is developing devices with new medical properties, and nanoparticles are being developed specifically for medical applications. A helpful definition of nanomedicine is 'the use of nanoscale or nanostructured materials in medicine that according to their structure have unique medical effects' (*ibid.*, p. 1,212).

Nanoparticles have unique properties that make them very attractive in medicine. They can cross biological barriers that are typically difficult for drugs to get through; they can accumulate preferentially in cancer cells, and they can increase the solubility of drugs and other molecules that otherwise dissolve poorly (*ibid.*). The unique magnetic properties of some nanoparticles are opening up completely new approaches to treating diseases.

As with all nanotechnology, a wide variety of products and devices is included within nanomedicine, and the usual broad division between futuristic and normal nanotechnology exists. Futuristic applications like nanobots will be examined in the next chapter, while this chapter will focus, for the most part, on 'normal nanomedicine'. Some of these available applications of nanomedicine will be described before looking at their ethical implications.

IMPROVED DRUG DELIVERY

By far the most dominant application of nanomedicine to date is improved drug delivery systems. Drug solubility is a significant problem, with estimates that about 40 percent of typical small-molecule drugs in development are 'brickdust candidates' (*ibid.*, p. 1,216) – in other words, no matter how effective or safe they might be, these drugs are as insoluble as brickdust, and about as valuable. By making them more water-soluble, they can at least be tested to determine if they are effective and safe enough to enter clinical trials. When drugs are poorly water soluble, higher doses are required to get enough drug into the body, and this can increase the risk of side-effects.

One strategy takes advantage of the large surface-to-volume ratio of nanoparticles. Substances dissolve as the surface molecules enter solution. The larger a particle, the longer it can take for it to dissolve completely. The Irish pharmaceutical company Elan has developed a patented procedure called NanoCrystal Technology. Machines grind the wet drug until its particles are 80 to 400 nm in diameter (Elan, 2007). These particles form a colloid that acts like a solution. The procedure has led to approval of four new formulations of previously approved drugs, and several others are in various stages of clinical testing. The nanoscale formulations can be taken orally, do not need to be taken with food, are absorbed better, can work faster, and may allow lower doses to be used. They have also generated over $1.5 billion in worldwide sales for their manufacturers.

Another general approach to improving drug delivery is to attach groups to the drug molecule that increase its overall water solubility. The resulting molecule can have nanoscale dimensions. For example, interferons are a class of naturally occurring proteins produced by the immune system to combat viruses. They can also be used as drugs, usually to fight cancer or viral infections. A nanomedicine version of interferon has been developed by attaching chains of polyethylene glycol (PEG). The product is called PEGylated interferon and its molecules have dimensions in the nanoscale. While its size puts it within the domain of nanomedicine, the chains are attached using conventional drug-design chemistry, pointing to the lack of clear boundaries between nanomedicine and conventional medicine.

PEGylated interferon is more effective than standard interferon, and yet it has comparable side-effects (Thomas and Foster, 2007). One major advantage for patients is that it can be given in one weekly injection instead of the three injections per week required for standard interferon. However, production costs are substantially higher, and larger quantities of the drug must be given. While the initial costs of treatment are increased, in the long run the PEGylated drug is calculated to be more cost-effective. This is an example of a 'nanotechnology-enhanced' pharmaceutical, where the characteristics of a previously approved drug are improved by nanotechnology (Wagner *et al.*, 2006, p. 1,213).

Another example is Abraxane, a nanoscale combination of albumen (another naturally occurring protein) and paclitaxel, an established anti-cancer drug (Bawarski *et al.*, 2008). Paclitaxel is poorly water soluble, requiring the addition of another solvent to give the traditional injected form, Taxol. This solvent causes hypersensitivity reactions, which require both gradual administration of the drug over three hours and other drugs to counteract the side-effects. Abraxane nanoparticles are about 100 nm in size, avoid the use of the toxic solvent and other drugs, and the full dose can be delivered in 30 minutes.

Nanotechnology can improve drug delivery in other ways, such as through liposome mediated transfer. Liposomes have been produced for over 40 years and can be regarded as 'one of the earliest forms of nanomedicine' (Sun *et al.*, 2008, p. 1,256). Liposomes are like tiny spherical cells. They vary in size, but are almost always less than 400 nm in diameter (Bawarski *et al.*, 2008). They contain a membrane made of lipids (fat molecules) which encloses an aqueous interior. Various kinds of drug can be encapsulated within liposomes, either water-soluble ones in the aqueous interior or water-insoluble ones in the lipid membrane. Liposomes can act as transporters of drugs, but they can also change a drug's biological effects. For example, the side-effects of the anti-cancer drug doxorubicin (brand name, Adriamycin) are serious and have limited its use. Doxil, a liposome-encapsulated formulation of doxorubicin, is much less toxic to the heart, and was one of the first nanomedicines to enter clinical practice (Bawarski *et al.*, 2008). However, problems with liposome stability and batch-to-batch reproducibility, as well as the cost of making liposomes of pharmaceutical-grade purity has slowed their introduction and kept them expensive (Wagner *et al.*, 2006).

A related nanostructure is the polymeric micelle, typically 50 nm in diameter, and used to increase water solubility (Bawarski *et al.*, 2008). They work much as soap coats water-insoluble dirt to make it washable in water. Micelle-enclosed drugs are also protected from the body's methods of eliminating them, allowing them to act as slow-release formulations. Several micelle formulations are already being used in clinical practice, and other nanostructures are being developed. In this way, normal nanotechnology is said to offer 'endless opportunities' for improving pharmaceuticals (Bawarski *et al.*, 2008, p. 274).

TARGETED DRUG DELIVERY

In discussions with the UK public, the improvement of targeted drug delivery was seen as one of the most positive applications of nanomedicine – much more so than discovering new drugs (Bhattachary *et al.*, 2008). Drug development is still influenced by the 'magic bullet' concept first promoted by the German medical researcher Paul Ehrlich (1854–1915). According to this concept, an ideal drug should target a specific part of the body, cure the problem and have no effects anywhere else. Two problems plague the development of so-called magic bullets. One is limited drug solubility, and the second is that many drugs affect different parts of the body, increasing the risk of side-effects. Nanotechnology is addressing both these factors.

Dendrimers are a complex type of nanoparticle receiving much attention within nanomedicine (Bawarski *et al.*, 2008). Dendrimers have a core unit onto which a number of branching chains are attached, giving a tree-like structure, with the name coming from *dendron*, the Greek for tree (Edwards, 2006). In three dimensions, these structures are spheres, with numerous cavities into which drugs, diagnostic agents or other molecules can be inserted for delivery to particular tissues.

Dendrimers can be targeted to specific tissues. Molecules that circulate in the blood often attach to specific sites on cells, where they can trigger an effect. For example, insulin circulating in the blood will attach to insulin receptors on cell surfaces, thereby triggering the cell to pull glucose in from the blood. The cells in different tissues have different receptors which may provide a way of targeting cells for particular drugs. For example, chemical groups or nanoparticles can be attached to the ends of dendrimer branches. These surface groups are chosen so that they interact with specific receptors found on targeted cells and tissues. Since one dendrimer can have several reactive sites on its surface, the biological effect can be greatly enhanced compared to a single molecule. The first dendrimer-based drug was approved for clinical research in 2003 as a vaginal microbicide designed to prevent transmission of sexually transmitted diseases including HIV (Bawarski *et al.*, 2008). The product, Vivagel, remains in clinical trials (AIDS Alert, 2008).

Nanoparticles can be targeted by attaching groups that carry out specific functions, leading to what are called 'functionalized' surfaces. For example, antibodies can be attached that recognize specific cancer cells allowing 'active targeting' which 'has revolutionized nanomedicine' (Bawarski *et al.*, 2008, p. 279). Nanoparticles in the body become coated with proteins which allow them to enter cells along specific channels (Lynch and Dawson, 2008). Such a nanoparticle-protein 'corona' is opening exciting possibilities of going beyond targeting specific cells to directing nanoparticles to specific compartments within cells. Much remains to be done here, but nanotechnology is making Ehrich's magic bullet approach look much more feasible.

NANOPARTICLE DRUGS

Nanoparticles can act as drugs themselves, and there is much interest in their use against cancer. Killing cancer cells while minimizing damage to normal cells is a major challenge, with strategies based on key differences between the two types of cells. Cancer cells grow rapidly, as do blood vessels around the cancer, but lymph vessels develop slowly (Bawarski *et al.*, 2008). The lymph system is involved in immune reactions and drainage from tissues. In addition, cancer cells tend to be more porous to nanoparticles than normal cells. Thus, nanoparticles are quickly delivered to cancer cells via the blood, easily enter the porous cancer cells and are removed more slowly because of poor lymphatic drainage. The overall result is that nanoparticles accumulate in cancer cells where they can selectively have their therapeutic impact.

Silver has long been known to kill bacteria. Bulk silver does not dissolve well, and while this is good for those who like to wear their silver in the rain, it is bad for those looking for an antibacterial agent. However, silver is a textbook example of a material with significantly different properties in bulk compared to nanoscale (Edwards, 2006). Nanoparticles of silver enter bacteria rapidly and interfere with several biochemical processes, making it less likely that bacteria will develop resistance. Acticoat is a nanosilver dressing for serious burns and wounds which is effective at preventing infections and can be left on the wound for a week, instead of other dressings which must be changed frequently, causing much pain.

More complex nanoparticles are also being developed, such as anti-cancer drugs that work via nanoparticle-based magnetic hyperthermia (Wagner *et al.*, 2006). Magnetic nanoparticles can be made from iron oxide and coated with other compounds so they are taken up selectively by specific tissues or cancer cells. The nanoparticles can also be injected directly into tumours, especially those which cannot be removed surgically. When exposed to an external alternating magnetic field, magnetic nanoparticles generate enough heat to kill the cancer cells in which they have accumulated (MagForce, 2009). As these same magnetic nanoparticles can be used as contrast agents to produce highly detailed MRI images, it is hoped that, in the future, tumours will be precisely irradiated, guided by real-time images from MRI (Hadjipanayis *et al.*, 2008).

Gold nanoshells are another type of fabricated nanoparticle (Liu *et al.*, 2008). A silica bead of about 100 nm can be coated with gold nanoparticles, which provides the seed layer for growth of a continuous gold coating. In nanoparticle hyperthermia, the gold nanoshells are injected into tumours or otherwise accumulate there. A laser beam of infra-red radiation is directed at the nanoshells and will pass through healthy tissue without affecting it. When the infra-red beam hits the nanoshells, the gold nanoparticles absorb energy and get warmer. An increase of only a few degrees is enough to kill cancer cells without harming the surrounding healthy cells.

Many other developments are taking place in nanomedicine, as part of what has been called the current 'nanotechnology gold rush' (Roukes, 2007, p. 4). While there have been many positive developments in drug delivery, much remains unknown about the potential toxicities of nanoparticles. A degree of caution is still warranted, though nanomedicine holds out great promise for patients. Richard Smalley commented on his own struggle with leukaemia prior to his death in 2005 at age 62.

> Twenty years ago, without even this crude chemotherapy, I would already be dead. But 20 years from now, I am confident we will no longer have to use this blunt tool. By then, nanotechnology will have given us specially engineered drugs, which are nanoscale cancer-seeking missiles . . . I may not live to see it. But, with your help, I am confident it will happen. Cancer – at least the type that I have – will be a thing of the past.
>
> (Smalley, 1999, quoted in Halford, 2006, pp. 14–15)

NANOMEDICAL ETHICS

On one level, the ethical issues with normal nanomedicine are similar to those arising with all forms of medicine and medical research. New products should be tested and evaluated for effectiveness and safety; clinical research should be conducted according to widely accepted ethical guidelines (Emanuel *et al.*, 2000); approved treatments should be offered, along with the information necessary for patients to make informed decisions about different treatments. Given the overlap, it has been assumed that nanomedicine does not raise new ethical issues for medicine. Thus the US National Institutes of Health (NIH) and the National Cancer Institute (NCI) have decided that within the funding available for nanomedicine, none will be allocated to the ethical and social issues (Khushf, 2007).

Nanomedicine does raise some of the general nanoethics issues, including the importance of balancing reality and hype. While developments in nanotechnology are said to be 'revolutionizing the future of medicine'; currently 'most of these visions are hypothetical' (Bawarski *et al.*, 2008, pp. 273, 280). This can become ethically problematic if the benefits blind us to the pitfalls, as some allege is already happening (*ibid.*). Little attention has been paid, or research devoted, to the health of those currently working with nanoparticles or their environmental impact. Of the billions already spent on nanotechnology research, only 1 percent has gone towards occupational health and safety research (*ibid.*, p. 280). As discussed in Chapter 5, lessons must be learned from the way those working with asbestos, PCBs and other new chemicals were early victims of ignorance and uncertainty regarding the effects of new products (Harremoës *et al.*, 2001). However, 'up to now the FDA has not required special testing of products containing nanoparticles' (Wagner, 2006, p. 1,217). This is particularly problematic with nanoceuticals (nanotechnology-enabled dietary supplements) which people consume but which do not require approval before going on the US market (Erickson, 2009).

The unique properties of nanoparticles that allow them to target cancer cells may have other, unexpected effects. Nanoparticle-protein coronas that reach intracellular compartments may accumulate there. 'Although the use of nanotechnology in drug imaging, diagnosis, and cancer therapy may be beneficial, it may also cause unintentional human exposure with unknown health effects that can only be imagined at present' (Bawarski *et al.*, 2008, pp. 279–80). Some people are imagining these effects, and imagining the worst. Hence, what is needed is careful evaluation of the harms and benefits of each new product on a case-by-case basis.

ETHICAL ISSUES WITH NEW CLASSES OF DRUGS

As new medical products are developed, they must undergo clinical testing to establish their effectiveness and safety. In many ways, the types of effects (both positive and negative) to be expected from new conventional drugs are well known, and batteries of tests are put in place to check for well-known effects. Standard protocols have been developed over years of closely monitored drug trials. Some nanoparticle pharmaceuticals can be expected to work like conventional pharmaceuticals and to be adequately monitored in conventional trials.

But others might not.

The more unique and innovative a development in nanomedicine, the greater the possibility of unexpected outcomes. An event took place in London in 2006 which pointed to the importance of a precautionary approach with new classes of drugs. Eight healthy male volunteers were enrolled in a controlled study of a new experimental drug called TGN1412 (Suntharalingam *et al.*, 2006). Two received intravenous saline as a placebo, and six others received the test drug by IV. Roughly an hour after the experiment started, the volunteers getting TGN1412 developed back pain and severe headaches. Then came restlessness, nausea, vomiting or diarrhoea. They became very hot and started to shake violently. Some had severe and rapid swelling around their neck and head. Their blood pressure dropped dangerously low.

About five hours into the experiment, things seemed to stabilize. But then one man began to have difficulty breathing, and his organs started to fail. He was rushed to a nearby intensive care unit, and the others were brought there as a precaution. All eventually had multi-organ failure, with two requiring intubation and mechanical ventilation.

Two days later, the four patients less severely impacted began to recover. The others required several more days in the ICU, followed by months of follow-up care for the most severely impacted man. The long-term consequences remain unknown. Why the men reacted so differently to the same dose of drug is also not understood.

The clinical symptoms that the men suffered are known as a 'cytokine storm' (Suntharalingam *et al.*, 2006). Cytokines are a group of biochemicals that trigger the immune system and other parts of the body (Expert Scientific Group, 2006). TGN1412 was a completely new class of drug called a superagonist monoclonal antibody. The drug was expected to be useful against diseases like rheumatoid arthritis or certain types of leukaemia that involve parts of the immune system (Bhogal and Combes, 2007).

Although this was the first time such drugs were given to humans, many other treatments using antibodies and antibody-related pharmaceuticals exist. These have been tested extensively and used widely, without causing problems like these. They often work better and more safely than older treatments. The reasons for the TGN1412 problems are complex, but relate to the novel way it was designed to impact many different pathways in the body.

While TGN1412 does not specifically fall within the field of nanomedicine, this trial serves to highlight important ethical issues that nanomedicine must address as its treatments enter clinical trials. The researchers did many laboratory and animal studies first, they followed standard practice in designing the TGN1412 trial, and it was approved by all the appropriate UK regulatory agencies and ethics committees. The participants were informed of the known potential risks, and all gave informed consent. Yet the tragedy occurred.

Application of the precautionary principle may have lessened or prevented some of the suffering caused by this trial. The generally accepted protocols for new conventional drugs did not protect these healthy volunteers, and they may not adequately protect the participants in early clinical trials of new types of nanomedicines. The British body set up to investigate the TGN1412 research concluded that additional caution is needed when testing biological molecules with a novel mechanism of action, new agents with a high degree of species specificity, or new agents targeting the immune system (Expert Scientific Group, 2006). Many nanomedicines will fall into one or more of these categories.

Prior to this incident, recommendations were made by a US research ethics body that certain types of research needed to have particularly careful scrutiny (Levine *et al.*, 2004). To warrant this precaution, such research would involve: 1) translating new scientific advances into human beings using interventions that are novel, irreversible, or both; 2) known or credible risk for significant harm without benefit to the participants; and 3) ethical questions about the research itself for which there is no consensus or there are conflicting or ambiguous guidelines. The TGN1412 trial satisfied all three criteria (Levine and Sugarman, 2006). Many new interventions being developed within nanomedicine will also and should be subjected to more precautionary investigations.

In the TGN1412 trial, the six participants received the drug within 5 to 10 minutes of one another. A more precautionary approach would recommend observing each participant for a longer period of time before giving the drug to the next participant. In addition, each patient received the drug over a few minutes, while a more precautionary approach would give the doses by slow infusion over hours. The dose of drug selected was 500 times less than that given to animals. While this might appear reasonably cautious, it was clearly mistaken in this case.

In medical research, doses, risk estimations and cohort sizes can be rough estimates based on experience or standard practice. A more precautionary approach would require careful statistical calculations of all relevant doses and quantities (Royal Statistical Society, 2007). However, part of what makes immune molecules attractive in nanomedicine is the minute quantities required. In some cases, just one molecule of a biological substance may be enough to trigger a response in the body. This two-edged sword may also bring unique risks and uncertainty.

Much debate has occurred over whether the TGN1412 researchers could have, or should have, realized that a cytokine storm might have occurred. Precaution is necessary when there is significant uncertainty about the transferability of preclinical information to humans. Some information has come to light that a few animals did have adverse effects during earlier research (Hama, 2006). These side-effects were explained away because of what was known about the way conventional drugs act. But with a completely new class of drugs, and unknown potential for good or harm, the anomalous animal results should have been investigated more thoroughly.

These types of biopharmaceuticals raise the controversial, expensive and difficult issue of animal experimentation. The safest approach for humans would involve testing the drugs extensively in animals, including on primates as they are most similar to humans. However, such experiments raise serious ethical issues with strong opinions on both sides. Researchers ascribe to the 3R principles of replacement, refinement and reduction of animals in research (NC3Rs, n.d.). The lack of alternatives to animal testing for new drugs that could relieve much human suffering creates major ethical dilemmas, which need very careful examination.

The precautionary principle applied to new research does not necessarily require that the research be abandoned, unless evidence demonstrates that this is reasonable. A precautionary approach might have abandoned the TGN1412 trial before it started, if other tests pointed to the high risk of a cytokine storm. Because of the drug's unique properties, the problems might not have appeared until TGN1412 was administered to humans. If the drug had been given according to the precautionary principle, fewer people might have suffered or the extent of the problems might have been reduced. However, we cannot be sure of avoiding all harms because of individual variation in our immune systems.

A precautionary approach needs to be taken with many developments in nanomedicine because, by definition, the first trials with new drugs or devices will involve much uncertainty. The principles discussed in Chapter 5 can guide the specific requirements for particular studies. As more scientific information becomes available, a more usual risk-benefit analysis can guide subsequent decisions. Thus, the precautionary principle need not be a hindrance to scientific and technological development, but can require 'more and better science' (Martuzzi, 2007, p. 570).

While a precautionary approach should help protect patients and the environment, it may also add costs. It will tend to slow down some areas of research and development, which can be viewed as a hindrance. The rapid pace of research can be gauged from patent filings. During the 1990s, worldwide patent filings in nanomedicine gradually increased, with a couple of hundred per year. In 2001, the application rate jumped dramatically, and has continued to increase to a point where more than 2000 were filed in 2003 – a ten-fold increase in ten years (Wagner *et al.*, 2006).

While the pace of discovery is exciting from a research and development perspective, it raises concerns about how well safety, regulatory and ethical oversight can be maintained. The economic pressure to see a return on the massive investment in nanomedicine puts pressure on researchers and manufacturers to push through clinical testing and to bring products on to the market. While this is not incompatible with due precaution, it does create forces that will require strong ethical resolve to resist.

DIAGNOSTICS IN NANOMEDICINE

An area of significant research in nanotechnology is the development of what is being called 'lab-on-a-chip' or 'point of care' technology. Current diagnostic tests often have to be sent to medical labs for analysis, taking time and involving a number of medical visits. The idea behind lab-on-a-chip technology is to shrink all the processes so that a drop of blood can be analysed quickly in a hand-held device (Leary *et al.*, 2006a). The challenge is to separate and purify nanogram quantities of materials from drops of blood and analyse nanolitre volumes (Wen *et al.*, 2008). Microfluidics is the term used for some of these developments which incorporate various developments in nanotechnology.

Lab-on-a-chip devices are part of the trend towards more personalized medicine and an emphasis on early diagnosis. The aim is to allow patients to receive results immediately from their physicians, or even to use the devices themselves at home. Such technology may also be useful in isolated regions or in countries which do not have medical laboratory infrastructure. James Bond in *Casino Royale* (2006) demonstrated the advantages of rapid diagnosis using the 'point of care' devices in his car after he had been poisoned. Lab-on-a-chip results could be displayed on the devices, or could be transmitted wirelessly to healthcare centres. In research with the UK general public, such devices were viewed as potentially the most promising development in nanomedicine (Bhattachary *et al.*, 2008).

However, the psychological impact of constant monitoring (mentioned in Chapter 4) is a concern: professional support would be important, especially if more serious conditions or ones for which treatment was not available or difficult to obtain were diagnosed (Kearns *et al.*, 2009); and the devices would also need to be carefully tested for accuracy and reliability. They raise specific ethical concerns when used for genetic diagnosis, to be discussed in Chapter 9.

NANOTECHNOLOGY SURGERY

The instruments of nanotechnology can also be used in medicine. Scanning tunnelling microscopy was a crucial breakthrough allowing researchers to 'see' things at the nanometre level. The atomic force microscope (AFM) can scan surfaces with a 3-micron pyramid probe with a tip about 30 nm wide (Ebbesen and Jensen, 2006). As living cells tend to be 5 to 10 microns wide (5,000 to 10,000 nm), researchers have modified AFM probes to make ultrathin nanoneedles, and, for the first time, an instrument has penetrated the nucleus of a living cell without causing fatal damage (Obataya *et al.*, 2005). Such techniques may allow insertion or removal of specific components within the cell, either as a form of therapy or for further analysis. Much interest has been expressed in using this method to deliver single genes into the nucleus of a living cell (Han *et al.*, 2008). However, the instrument can only be used on isolated cells, not on those inside the body. To date it has been used on living cells in the lab, though it may find clinical application with *in vitro* embryos prior to implantation.

Another approach to nanosurgery involves extremely short-burst lasers. Medicine has used lasers for decades, most noticeably in eye surgery. However, they can produce excessive heat that damages tissue around the surgical site. Femtosecond laser surgery uses pulses so short that nanoscale incisions can be made without heat damage (Leary *et al.*, 2006b). A femtosecond is one millionth of a nanosecond. The laser energy is so intense that objects are vaporized, leaving no residual particles to cause problems in the body. Femtosecond laser surgery has been used to cut individual chromosomes within cells, and this could lead to applications in gene therapy. In another experiment, one mitochondrion in a living cell was destroyed while leaving others unaffected, even though these were only a few hundred nanometres apart (*ibid.*).

Carbon nanotubes have been used to make 'nanotweezers', 'nanoscissors' and other instruments that could permit actual nanosurgery. These tools could eventually be used to manipulate and modify many biological structures that exist within living cells (*ibid.*). Nanotechnology is moving closer to a *Fantastic Voyage* scenario, where cellular damage can be repaired on a subcellular level. Here is where normal nanomedicine comes close to futuristic nanotechnology and its nanobots, but we will leave that discussion until the next chapter.

NANO-ENABLED IMPLANTS

Science fiction accounts of nanomedicine often focus on nanotechnology implants. In *The Diamond Age*, one of the main characters, Bud, has a 'testosterone pump' implanted in his forearm, and this leaves him looking muscular 'except you didn't have to actually do anything and you never got sweaty' (Stephenson, 1995, p. 3). Implants are tiny devices placed inside human bodies to regulate a biological function or to deliver a drug, hormone or other substance (Leary *et al.*, 2006b).

Those working within normal nanotechnology have a view different to that presented in futuristic accounts. 'In fact, active implants is the medical sector for which experts predict the least impact of nanotechnology' (Wagner *et al.*, 2006, p. 1,213). Nanotechnology may play an enabling role in developing smaller electronics and coatings to make implants more compatible with tissues and to allow better connectivity between implants and cells. Thus implants are likely to be nano-enabled, and not nano-scale. They may be controlled by nanochips to deliver precisely adjusted dosages, thus benefiting patients with many different conditions. Over 200 proteins and peptides, including widely used products like insulin and growth hormone, are approved for medical use in the USA. Almost all must be given by injection because, like other proteins, they would be rapidly digested if taken orally. Implanted devices could eliminate the need to inject these proteins.

Implants can raise significant ethical issues, however. One that delivers insulin to a diabetic patient in a precisely controlled and individualized regime would be important and ethical. But the same devices could easily be adapted to regulate hormone levels to keep someone's mood stable (or unstable, depending on one's preference). Bud's testosterone pump raises ethical questions about non-therapeutic uses. Some day, surgeons might be expected to submit to having implants with drugs which could help them to concentrate or keep their hands from quivering. How much control do we want to give to the settings on the nanochip? Will someone be holding a remote control to keep our behaviour 'normal'? The path from ethical medical practice to science fiction scenarios is smooth and plausible, and normal nanotechnology is starting us down that road. The issues are complex and challenging, and will be the focus of the next chapter.

REPLACING BODY PARTS WITH NANO PARTS

The scarcity of organs for transplantation has inspired numerous science fiction stories like *Spares* (Smith, 1996), *Never Let Me Go* (Ishiguro, 2005) and *The Island* (2005). The great need for organs can appear to justify a host of ethically controversial technologies. Denzel Washington's hostage-taking approach in *John Q* (2002) used only current technology, but was just as ethically controversial. Nanotechnology may, in some instances, offer an alternative strategy. Rather than using human or animal organs, artificial organs may work as well or even better. This approach comes with its own ethical issues, as some of the devices involve the first direct brain-machine connections.

Cochlear implants were the first devices to offer restoration of one of the five basic senses. First developed in the 1970s, their development has been aided by recent advances in nanotechnology. Some people lose their hearing when the tiny hairs inside the inner ear are damaged and lost. A cochlear implant involves surgically placing a set of tiny electrodes deep inside the inner ear (Chorost, 2006). The electrodes stimulate various nerve fibres in the ear and send messages to the brain. A functioning auditory nerve is necessary for the implants to allow people to detect sounds. The electrodes are attached to a receiver that is implanted under the scalp along with a magnet. This holds a removable wireless transmitter to the scalp, with the transmitter attached, via a belt-held processor, to a microphone that looks like a hearing aid.

The restorative value of the cochlear implant is immense. The deaf hear. But the devices have their limits. Hearing in quiet surroundings can be improved, but hearing amidst noise is very limited. The quality of music is described as being very poor (Clark, 2008). Carbon nanotubes are being examined to improve the quality of hearing via cochlear implants. Polymer-coated nanotubes (with specific proteins incorporated) are being tested both to prevent further degradation of ear cells and to transmit sound signals to the brain. In addition, carbon nanotubes may have a role in promoting electrical stimulation of nerves, and this may help in future developments. Advances in nanotechnology are predicted to lead to a second generation of cochlear implants which are part of a new discipline called 'medical bionics' (*ibid.*, p. 686). The reference is to the 1970s television show *The Six Million Dollar Man*. Completely different approaches to restoring hearing are being pursued using other types of nanoparticles and brainstem implants (Edwards, 2006).

Some commentators are concerned that these cochlear implants both stigmatize deaf people and put pressure on them to obtain implants. Another, deeper issue is how implants impact upon a person's self-identity. Michael Chorost's autobiography describes the triumphs and tribulations, the exhilaration and depression that go with receiving a cochlear implant. Although changed for the paperback edition, the original subtitle of his book was: *How becoming part computer made me more human* (Chorost, 2006). His journey with the implant was a struggle. 'And now I am becoming something else: not *in*human, not *post*human, but *differently* human' (*ibid.*, p. 33). He uses science fiction to explore the possibilities. He concludes he is not like Terminator, as Schwarzenegger's character was a robot. He worries that he is undergoing a Borg assimilation, and losing control of himself, 'perceiving the world by a programmer's logic and rules instead of the ones biology and evolution gave' him (*ibid.*, p. 9). He concludes that because of this external control he is 'a *cybernetic organism*, an organic creature whose body is controlled by algorithmic rules' (*ibid.*, p. 71). In spite of the ups and downs, the fact that his batteries run out and he sometimes gets a bad signal, he is grateful for the gift of hearing and all that it brings. 'I can just walk out my door and encounter the world whole and full. Rebuilt' (*ibid.*, p. 195). Not everyone receiving an implant is impacted this deeply, but his story points to the ethical responsibility of providing psychological support for recipients.

Much has also been written on technology's potential to cause alienation. Chorost states that his cochlear implant did not have this effect on him, but rather he felt 'more connected to the world post-activation, rather than less' (*ibid.*, p. 188). Others have not had as positive an experience. An older couple who received cochlear implants struggled to adapt. In the television documentary *Hear and Now* (2007), the husband concluded that his life was enhanced much more by an electronic device that allowed text communication via the telephone. This gave him a connection with other people, while the implant, he said, only allowed him to hear. Technological replacement parts may be beneficial and technically successful, but their impact on the person can be very complicated.

Meanwhile, other artificial replacement organs are being developed. An artificial vision system has allowed a blind man to move around without bumping into objects (Dobelle, 2000). Interviews with blind people had revealed that mobility was generally more important to them than being able to read (Friedman and Kolff, 2000). The system consisted of a miniature digital video camera attached to the blind person's sunglasses. Electrodes were implanted into the blind person so they can stimulate the visual cortex of the brain. A small post protrudes from the person's head to attach leads coming from a portable computer. This interfaces with the camera so that the blind person 'sees' an array of lights which outline the objects in front of the person (Dobelle, 2000). Alternatively, the digital feed from the camera can be replaced by that of a television or computer so that the blind person can visualize content from the TV or internet. The image quality was very poor, like having a computer screen with very few pixels. Nanoscale engineering is expected to allow fabrication of electrode arrays with much higher resolution.

The technology to restore sight had advanced by 2008 to where it was being described as a 'bionic eye'. In NBC's unsuccessful and short-lived 2007 remake of *The Bionic Woman*, the nanotechnology-enhanced Bionic Woman was not able to capture a new generation of viewers, even with her fully zoomable bionic eye. Back in today's world, eye surgeons working on restoring sight to the blind acknowledge that their approach 'is straight out of science fiction' (Smith, 2008). The device is very similar to those already described, with a camera, implanted receiver and electrodes, and an external iPod-size transmitter. People see fuzzy blocks of light and dark, just sufficient to identify objects and to manoeuvre without bumping into things. As with cochlear implants, the brain must learn to process the visual signals it receives. This can be difficult depending on whether recipients were always blind, or how long they had been blind (Alda, 2005). Researchers anticipate that the devices will improve to the point where blind people will be able to read. The devices require an intact optic nerve, so they will only help some blind people.

Even more dramatic improvements in restoring eyesight may result from another breakthrough in electronics (Highfield, 2008). Conventional digital imaging is built upon rigid semiconductor materials. This restricts their usefulness to flat surfaces, which is far from optimal with curved surfaces like the eye. Researchers have developed a way to make 'stretchable electronics' using a flexible mesh of wire-connected sensors and other nanotechnology-based developments (Ko *et al.*, 2008). The researchers developed a working camera with a hemispherical shape, but with relatively low resolution. The sharpness of the images was significantly better than those taken with a simple, planar camera. The curved camera gives hope that an artificial eye may be feasible soon.

Devices like cochlear implants and artificial eyes require changes in the brain to process the electronic signals. These can take months or years, and may never be perfect. Young people adapt more easily which is already raising ethical concerns. Parents may feel pressured to put implants with life-long consequences into minors before they can give proper consent. For implants with clear therapeutic purposes, this may be *relatively* uncontroversial. However, once developed, the devices will be easily adaptable for non-therapeutic purposes. Children may be given enhancement or experimental implants before they are adequately tested or before all the psychosocial implications are understood. Using implants for enhancement purposes raises serious ethical issues which will be discussed in the next chapter.

BEYOND TODAY

Stretchable electronics are anticipated to have many applications, including various bionic implants and robotic sensory skins (Someya, 2008). Electrodes could be shaped to any part of the brain or body. The curved camera was inspired by design features in animal eyes, leading to predictions of 'artificial insert-like compound eyes . . . [and] fish eyes that have a 360° field of view' (*ibid.*, p. 704). The connection with science fiction is clear. 'An electronic eye that works like the real thing foreshadows the development of a new generation of bionic eyes and other "cyborg" technology seen in the film "Terminator" and other Hollywood sci-fi movies' (Highfield, 2008). Connecting any part of the brain to an electronic device or external computer moves us closer to the cyborg era.

Once artificial eyes restore normal sight, very little further modification will be needed to enhance their capabilities. Just as with conventional cameras, lenses with zooming capacity will be developed. Artificial eyes that allow vision in the infrared or other ranges will be available with relatively minor adjustments to the basic model. The military applications are obvious. Indeed, 'military interests in R&D have shaped the goals of nanotechnologies from the very beginning' (Schummer, 2007a, p. 87). Serious discussions are needed regarding the goals and purposes of these developments.

> Enhanced physical strength, bullet-proof cloths, enhanced sensual capacities in the infrared or other ranges, enhanced mental capacities through brain-computer interfaces, and so on, all may well improve the military performance of soldiers. However, although these goals might appeal to some individuals, civil societies are built on different human values and different human qualities than those needed for military operations.
>
> (*ibid.*)

Meanwhile, work is ongoing with even more advanced types of brain implant (McGee and Maguire, 2007). In 1998, a patient with locked-in syndrome received a brain implant which allowed him to communicate with a computer by thinking (Headlam, 2000). The pain and anguish of people in this condition was graphically revealed in the autobiography and subsequent film *The Diving Bell and the Butterfly* (Bauby, 1997, and film, 2007). The author could move nothing other than blink an eye, by which he developed a code and dictated his book to his therapist. The benefit of a brain implant which allows more straightforward communication with people with this syndrome is clear. However, what they might tell us could create even more challenging ethical dilemmas, as revealed by Bauby. One of his first communications was that he wanted to die.

Other developments are occurring which could be combined with the medical developments. Using nanoscale fabrication methods, a contact lens has been made which superimposes virtual images onto whatever the eye sees (Hickey, 2008). The pop-up screens that science fiction's Terminator could see may not be too far away. The US military and NASA are actively pursuing research on brain-machine communication. The NASA research project called Extension of the Human Senses has a goal of developing 'alternative methods for human-machine interaction as applied to device control and human performance augmentation' (NASA, 2008). The plan is to use wearable devices that can transmit electrical signals to computers and other devices, with goals of silent communication and control of instruments or machines without physical connections. Other researchers are working on implanted electrodes that will allow people to control computers by thought alone (McGee, 2008). Where all of this is taking us, only science fiction seems to know.

NANOMEDICINE CHANGING MEDICINE

In some ways, the vision for nanomedicine is presented as if the doctor–patient relationship will remain the same, with nanomedicine providing more and better tools and treatments. Yet more careful reflection reveals that nanomedicine would require a radical transformation of medicine and the doctor-patient relationship. Nanotechnology may give doctors:

> . . . a hand-held device that will be able to analyze a fraction of a droplet of blood for 1000 or more proteins and these will be a window into health and disease. This will be done twice a year. The information will be fed into a cell phone and then to a server, and then it will be analyzed and the patient and the physician will get an email that says, 'You are fine; do this again in 6 months,' Or, 'you should see an oncologist.'
>
> (Hood, quoted in Khushf, 2007, p. 517)

This scenario is rapidly approaching, with lab-on-a-chip devices. These could alert people to changes in various biomarkers long before damage or symptoms appear. However, it could also move much of the difficult, complex and individualized clinical judgment to a remote computer. The physician would only take the blood, select the tests (assuming some selection is needed), and make sure the computer connection is working. Some might question why a physician would be needed here at all. The broader impact of separating medical diagnosis from the involvement of any healthcare practitioner must be carefully considered (Kearns *et al.*, 2009). Good reasons to support this separation can be envisioned, such as eliminating costly, time-consuming appointments and protecting patient privacy. But other consequences may not be so helpful.

Narratives can provide an opportunity to imagine the consequences of such changes. Some of these were explored in Chapter 3 when discussing *The Island* (2005). The film also suggests an important balancing point with technological diagnosis. Lincoln Six Echo's morning diagnostic test revealed a high sodium level which meant he was not allowed bacon for breakfast. He then does everything he can to get some bacon! Today, we don't have hand-held devices testing a thousand proteins. More primitive technologies, such as bathroom scales, no-smoking signs and TV ads, inform people about various health-promoting choices they can make. Many do not make these choices, or find them very difficult to implement.

Many people's health today is negatively impacted, not because they don't have clear diagnostic indicators, but because they struggle to make necessary lifestyle changes. While some diseases need better diagnostic technology, many do not. Overemphasising new technology to diagnose more conditions may simultaneously mean neglecting research into why people do not live as healthily as they should or even want to themselves. This is another area where technological determinism comes up short. Developing new technology can be more straightforward than developing strategies to help people adopt healthy behaviours, but without those strategies, the emails from the lab-on-a-chip computer will do no good. Someone may need to sit and talk with a doctor, nurse or someone else about why the lifestyle changes are so difficult to make. The move to lab-on-a-chip diagnostics may make those encounters less likely if it simultaneously leads to remote, depersonalized medicine.

At the same time as medicine becomes less personal, personal life is becoming more medicalized. Diagnostic devices are increasingly focused on risk, not disease. Some tests reveal if someone has an infection or cancer. Many others, like tests for cholesterol, blood pressure, blood glucose and a growing number of genetic tests, give readings that relate to the risk of developing disease. This information can be very important, especially if it helps prevent disease. But it can leave people more inclined to seek treatment, when they really need to focus on lifestyle changes. Decisions in this area are best handled between patients and their doctors, but appointments may be more difficult to arrange.

All of this is relevant to nanomedicine because decisions are being made about where to expend huge amounts of research monies. If new diagnostic devices inform us primarily about our *risks* for different diseases, but this is interpreted to mean that we *have* diseases, this will generate increased pressure to provide treatments, maybe unnecessary ones. More areas of our lives may become medicalized, both by diagnosis and treatment. This will create pressure to put more resources into risk reduction strategies, while many people with pathological diseases may be unable to access or afford treatments for existing diseases with ongoing pain and suffering. This doesn't even consider those in developing countries whom we discussed in Chapter 6.

Nanomedicine that seeks to treat illness and disease is based on the ethical mandate to heal the sick and relieve pain and suffering. Diagnosis of disease is an important part of that enterprise. Determination of risk of disease becomes a little less clear-cut, but can reduce illness if addressed in a multi-dimensional way. Nanomedicine that is not directed at illness but seeks to enhance the human condition is ethically more problematic. Developments in nanotechnology are making medical enhancement more feasible. For that reason we will devote the next chapter to this topic.

8 Enhancement: Becoming Better than Healthy

Medical developments with normal nanotechnology have the potential to greatly increase our capacity to prevent, diagnose and treat illness and disease. As their safety and effectiveness are established, they will contribute to the traditional goals of medicines curing and alleviating suffering. They also have the potential to significantly enhance human capacities. Futuristic nanotechnology is immersed in enhancement with nanobots providing the most dramatic examples. The vision is that nano-scale robotic devices will roam our bodies, monitoring functions and repairing problems. These devices are usually described as being a few microns in size, similar to that of bacteria. This makes them larger than nanoscale, but they will be heavily dependent on nanotechnology.

NANOBOTS

Nanobots have been part of nanotechnology since its beginnings. Richard Feynman's speech described 'small machines' acting as a 'mechanical surgeon' (1959/1992, p. 64). In the Foreword to K. Eric Drexler's book, Marvin Minsky stated that Drexler's approach would lead to 'tiny devices that can travel along capillaries to enter and repair living cells' (1986/2006, p. 19). This would bring the ability to 'heal disease, reverse the ravages of age, or make our bodies speedier or stronger than before' (*ibid.*, p. 20). Although involving treatment and prevention of illnesses, nanobots are primarily a means to enhance the human body.

Robert Freitas is the most prolific promoter of nanobots. He has written about 'respirocytes' that will carry oxygen in the blood with greater efficiency than red blood cells; 'microbivores' that will search out and destroy invading organisms better than our immune system; 'surgical nanorobots' that will carry out surgery inside the body; and 'pharmacytes' that will deliver drugs to precisely those cells where they are needed (2005, 2006). He claims these 'will be fabricated . . . perhaps in the 2020s' (Freitas, 2005, pp. 243–4) as nanobots exemplify 'a continuing and inevitable technological evolution toward a device-oriented nanomedicine' (Freitas, 2006, p. 2,770). His vision is impressive.

> Future nanorobots equipped with operating instruments and mobility will be able to perform precise and refined intracellular surgeries which are beyond the capabilities of direct manipulation by the human hand. We envision biocompatible surgical nanorobots that can find and eliminate isolated cancerous cells, remove microvascular obstructions and recondition vascular endothelial cells, perform 'noninvasive' tissue and organ transplants, conduct molecular repairs on traumatized extracellular and intracellular structures, and even exchange new whole chromosomes for old ones inside individual living human cells.
>
> (Freitas, 2005, p. 245)

Some maintain that this vision is purely science fiction, but work is proceeding. Kathie Olsen, Chief Scientist at NASA, stated in a documentary that NASA was developing 'nano explorers' which would 'be able to detect, diagnose and treat disease' (Olsen, quoted in FirstScience.tv, 2007). The commentator stated that these would involve people swallowing 'millions of microscopic machines, tiny robots designed to patrol the body, seeking out disease at the earliest stages' (*ibid.*).

A review of nanobot research in 2007 concluded that their manufacture was beyond current capabilities, but that progress was occurring in areas required for their development. Particularly in diagnosis and imaging, early nanobots could give 'far more rapid, flexible and specific performance than is possible with today's larger devices' (Hogg, 2007, p. 72). Optimal operation will require about a billion nanobots per dose, along with human monitoring.

One of the challenges for nanobots will be their potential rejection by the body. In *Fantastic Voyage* (1966), part of the film's drama was built around the body's immune cells recognizing the submarine and subsequently attacking it. Given nanobots' similar size to bacteria, immune rejection is likely. Another concern is waste produced when nanobots break down foreign or damaged cells. However, Freitas is confident that nanobots will break down such items completely to sugars, amino acids and other harmless, natural compounds (Freitas, 2006). This adds another layer of complexity to their capabilities.

The degree of complexity described by Freitas is astounding. The nanobots he envisions will be required to have a power source, biocompatible coating, an injector for delivering drugs, an antenna for wireless communication, molecular sorting pumps, separate carrier bays for each drug or hormone, molecular recognition sites, a waste disposal system, and possibly other items, such as rotors for steering. Responses to criticisms often lead to yet another device being incorporated.

The environment necessary for people to accept nanobots might be easier to create than the devices themselves. If constant monitoring and increased medicalization of life become widely accepted, people may be prepared to accept internal monitoring. If hand-held devices send data to computers that produce automated treatment recommendations, the need for healthcare professionals may diminish. It may seem perfectly reasonable to then transfer all these functions to nanobots. They will monitor levels of biomarkers and carry drugs and hormones to correct imbalances. This will take us right to Freitas's nanomedical scenario. While it may currently seem imaginative and speculative, step-by-step and motivated by important goals, we may end up with nanobots running through our bodies monitoring and correcting all sorts of physiological measurements.

Freitas gives us a nanomedical utopia. Others point out how things could go wrong – badly wrong. In the novel *Chasm City*, humans reach the stage where medical nanobots are widespread. Nanobots called medichines infuse human bodies. But then a plague causes massive destruction.

> At the time of the plague's manifestation our society was supersaturated by trillions of tiny machines. They were our unthinking, uncomplaining servants, givers of life and shapers of matter, and yet we barely gave them a moment's thought. They swarmed tirelessly through our blood. They toiled ceaselessly in our cells.
>
> (Reynolds, 2002, p. 3)

In the novel, the precise cause of the plague is unknown. Its appearance does not require the violation of any scientific laws but is something that can be realistically envisioned given what we know about viruses.

> It was not quite a biological virus, not quite a software virus, but a strange and shifting chimera of the two. No pure strain of the plague has ever been isolated, but in its pure form it must resemble a kind of nano-machinery, analogous to the molecular-scale assemblers of our own medichine technology.
>
> (*ibid.*)

We cannot predict what scenario will result, but we can look at the beliefs and values underlying a future filled with nanobots. Central to that vision is the motivation to enhance human capabilities. Nanobots represent one extreme on the spectrum of enhancement technologies. Current therapies already exist that promote enhancement, not treatment. The ethical debate is already raging over whether medical enhancement is a legitimate goal for medicine. Although the ethical issue is not generated by nanotechnology, how we decide will strongly influence how nanomedicine develops.

NANOTECHNOLOGY ENHANCEMENT

The power of medicine and drugs to change human behaviour, and humans themselves, has increased, and nanotechnology will augment these powers dramatically. K. Eric Drexler claims that 'nanotechnology can help life spread beyond Earth . . . It can help mind emerge in machines . . . And it can let our minds renew and remake our bodies – a step without any parallel at all' (1986/2006, p. 89). The overall impact will be on the nature of medicine and its goals. 'Physicians using scalpels and drugs can no more repair cells than someone using only a pickax and a can of oil can repair a fine watch . . . Cell repair machines will change medicine at its foundations' (*ibid.*, pp. 251–2). He elaborates:

> The technology underlying cell repair systems will allow people to change their bodies in ways that range from the trivial to the amazing to the bizarre. Such changes have few obvious limits. Some people may shed human form as a caterpillar transforms itself to take to the air; others may bring plain humanity to a new perfection. Some people will simply cure their warts, ignore the new butterflies, and go fishing.
>
> (Drexler, 1986/2006, p. 467)

This type of world is explored in science fiction. One female character in *Chasm City* is nicknamed Zebra because her skin has been modified to look like that of a zebra. Going beyond tattoos and body piercings, wealthy humans in this future time change their bodies with the help of Mixmasters. In the past, 'their work had been discreet: correcting genetic abnormalities in newborns; ironing out inherited defects in supposedly pure clan lines' (Reynolds, 2001, p. 394). Later, Mixmasters provided genetic changes based on what people desired and could afford. People visited their parlours and looked over brochures to select their next makeover. The options were almost endless, providing 'Anything you weren't born with, or weren't meant to inherit . . . The Mixmasters rewire your DNA so that the changes happen naturally – or as close to naturally as makes no difference' (*ibid.*).

We are a long way from this type of genetic manipulation. But some want us to move in that direction, based on the same ethical principle: human autonomy. People in *Chasm City* freely choose available technology, so long as they can afford it. The novel explores the implications of accepting that autonomous people have the right to choose to do whatever they want with their bodies and lives. The case presented here by science fiction is extreme, but the issue is not whether this is feasible. The example should cause us to reflect on the values and desires revealed by the scenario and what they say about our values and dispositions towards currently available options.

An editorial in *The Lancet* noted that those who argue against using gene therapy for enhancement 'do not give adequate weight to the latitude that our society affords to citizens who wish to enhance their physical appearance or their health' (Miller, 1994, p. 317). An earlier editorial in *The Economist* had asked:

> But what of genes that might make a good body better, rather than make a bad one good? Should people be able to retrofit themselves with extra neurotransmitters, to enhance various mental powers? Or to change the colour of their skin? Or to help them run faster, or lift heavier weights? Yes, they should. Within some limits, people have a right to make what they want of their lives.
>
> (Anonymous, 1992, p. 11)

The limits suggested are: if the changes cause harm to others, or if people want to make themselves psychopaths. James D. Watson, winner of a Nobel Prize for co-discovering the helical structure of DNA, puts the matter as a simple question: 'If we could make better human beings by knowing how to add genes, why shouldn't we do it?' (Watson, quoted in Brave, 2003). He advocates developing gene therapy and pre-natal screening tests to prevent low intelligence on the basis that, 'If you really are stupid, I would call that a disease' (Watson, quoted in Henderson, 2003). He has responded to allegations that he wants to use genetics 'to produce pretty babies or perfect people' by saying,

'What's wrong with that? . . . It's as if there's something wrong with enhancements' (Watson, quoted in Abraham, 2002).

Bioethicist Dan Brock holds that, 'An individual's choices – of a specific life plan, and to enhance the capacities necessary for that plan of life – are exercises of self-determination, not impingements on another's self-determination' (Brock, 1998, p. 55). He goes on to state that if people believe a particular enhancement therapy will help promote their chosen life-plan, they should have the freedom to chose it, 'even when they do so foolishly and with consequences that they may later come to regret' (*ibid.*).

Today, we cannot choose to have a genetic makeover that changes our skin colour. But we can have a surgical makeover that changes the shape of our face or body. We can pursue a pharmaceutical makeover that changes our brain biochemistry. We could soon have a nanotechnological makeover that changes our sense perception or strength or some other characteristic. The question is whether nanotechnology will benefit human society if used for enhancement, rather than treatment. We will begin by looking at whether or not a distinction can be made between treatment and enhancement.

THE TREATMENT-ENHANCEMENT DISTINCTION

One approach to determining whether some uses of medical technology are ethical has been called the treatment-enhancement distinction. In bioethics, the term enhancement applies when medicine and biotechnology are used to improve 'the "normal" workings of the human body and psyche, to augment or improve their native capacities and performances' (President's Council on Bioethics, 2003b, p. 13). The treatment-enhancement distinction assumes the legitimate goals of medicine are to treat or prevent disease and restore people to normal health, while medicine should not be used to enhance or improve people beyond the normal.

For example, testosterone is a steroid that can be used in various ways. Within the treatment-enhancement distinction, using it to treat an illness or to restore people to normal testosterone levels would be viewed as ethically appropriate. Using it to help an athlete get stronger is viewed as enhancing the athlete beyond normal, and therefore unethical. Such usage is also criticized because the risks associated with using it are not counter-balanced by the benefits of relieving an underlying medical condition, and because the drug is banned by most sports organizations.

The treatment-enhancement distinction requires clarity on concepts such as 'health', 'disease', 'normal' and 'enhance', though this is often lacking. Traditionally, good health was seen as the absence of disease and illness, and medicine promoted health by treating or preventing disease. In response to concerns that this view was overly focused on the physical and biochemical aspects of health and disease, the WHO adopted a revised and more holistic definition:

> Health is a state of complete physical, mental and social well-being and not merely the absence of disease or infirmity. The enjoyment of the highest attainable standard of health is one of the fundamental rights of every human being without distinction of race, religion, political belief, economic, or social condition.
>
> (WHO, 1948)

In 1984, WHO adopted a resolution to expand the definition to 'a dynamic state of complete physical, mental, spiritual and social well-being', although the Constitution has not been revised to reflect this (Khayat, n.d.). As society has accepted this broader view, the belief has simultaneously developed that medicine should use its knowledge and technology to promote physical, mental, social and spiritual well-being. Growth in the use of enhancement therapies suggests this broader purpose has been widely adopted. Those who view biomedical enhancement as inappropriate tend to view 'medicine as devoted to healing';

while those who favour biomedical enhancement tend to view 'medicine more broadly as helping patients live better or achieve their goals' (Greely *et al.*, 2008, p. 704). The implications of each view are substantial, not only ethically but also financially.

The treatment-enhancement distinction is not without difficulties – some medical interventions lead to enhancement *and* treatment. A common example is vaccination, widely accepted as medical therapy to prevent an infection or disease. However, vaccines work by enhancing the immune system so that it works beyond what is normal (Bostrom and Roache, 2007). In the same way, a therapy that slows the ageing process could be seen as a treatment to reduce the risk of illness or disease, or as an enhancement of normal lifespan.

Another difficulty arises with determining the normal state. Various tests are used to diagnose attention-deficit/hyperactivity disorder (ADHD). Experts disagree on which scores fit within 'normal', and which should lead to a diagnosis. This issue is complicated in ADHD because what is 'treated' is a set of behaviours including inattentiveness, distractibility, hyperactivity and impulsiveness. The challenge is that 'these behaviors are found to some extent in most normal children at some time or another' (President's Council on Bioethics, 2003b, p. 74). Another complication is that drugs used to treat ADHD are effective in reducing inattention, distractibility and impulsivity in normal children. Since it is difficult to determine what constitutes normal behaviour, it is also difficult to distinguish between treatment and enhancement. Some conclude that people should therefore be free to determine for themselves what drugs they want to use or have their children take.

Those favouring enhancement therapies also point out that people already enhance themselves and their children in many ways, including pharmacologically. Dramatically increased use of Ritalin, Viagra, Prozac and other so-called 'lifestyle drugs' points to this acceptability. Such drugs have uncontroversial value for medical conditions, but they are increasingly being used for 'the sole purpose of increasing personal life quality and of attaining a current psycho-socially defined beauty ideal, whereby no medical need for the treatment exists' (Harth *et al.*, 2008, p. 141).

Another challenge for the treatment-enhancement distinction arises from the observation that people's pursuit of enhancement is often lauded when other means are used. Education can be viewed as a means of enhancing cognitive abilities, practice and training enhance sports performance, private tutoring enhances musical talents and sermons can enhance moral character. For cognitive enhancement, at one end of the spectrum are gene therapy, brain–computer interfaces, and implants with neural growth factors, while at the other end are day planners, Post-it notes, and organizing our offices. Since all of these are means to cognitive enhancement, the line between acceptable and unacceptable is hopelessly blurry and some say searching for one should be abandoned. The argument is that since the goal of cognitive enhancement is accepted as ethical, any freely chosen means towards that goal should be ethically acceptable. A group of prominent scientists wrote that cognitive enhancement drugs

> along with newer technologies such as brain stimulation and prosthetic brain chips, should be viewed in the same general category as education, good health habits, and information technology – ways that our uniquely innovative species tries to improve itself.
>
> (Greely *et al.*, 2008, p. 702)

Enhancement enthusiasts make the same argument for other capacities.

Enhancement, in terms of human improvement, is a goal that society accepts and values. Each form of enhancement can be pursued by various means, some medical. A clear distinction between treatment and enhancement is difficult to establish. There is a blurry, grey middle. Almost every area of human activity comes with a spectrum, but we do not thereby eliminate classifications: they have their uses and advantages, even if they do not settle every question. For example, in parenting, most people would agree that discipline is an important goal. Children are disciplined in many ways, including corporal punishment, a slap on the cheek, yelling, Ritalin, time-outs, removing privileges and others. Where discipline stops and abuse starts is not objectively apparent. The conclusion should not be to declare that the distinction is worthless or that everyone should decide for themselves. We can establish what is acceptable, what is not acceptable, and where we are unsure or accept different approaches. As we noted with the precautionary principle in Chapter 5, blurry boundaries can demand careful thinking, which may lead to more balanced positions.

In many ethical dilemmas, different conclusions can be traced to a reliance on different underlying ethical theories. Utilitarian theories focus on the consequences that bring about the greatest good for the greatest number of people. Arguments in this tradition would note that enhancement is a goal that many people accept, since education, parenting, the arts, religion and other pursuits help people improve their lives. Therefore, if nanotechnology can further enhance people and their capabilities, such developments are ethical.

Another approach to ethics involves deontological theories, which focus on duties and on respect for the person. Such approaches would note that even if the consequences are good, the ends do not justify the means. While enhancement may be an ethical goal, not all means of achieving enhancement are automatically ethical. One of the many places where society uses this approach is in medical research. The goal of developing new treatments is widely accepted as ethical, but every means to obtaining new treatments is not thereby justified. For example, research that ignores or denies informed consent would be ruled as unethical, regardless of how good the goals. Again, parents are expected to have a goal of disciplining their children, but this does not justify any means parents might adopt, simply because they view it as acceptable.

In this approach, all means to a good goal (or good end) are not immediately accepted as ethical. Also, a means that is acceptable in one situation is not immediately viewed as ethical in every situation. The goal of athletic performance is good, but taking testosterone and training hard are not ethically equivalent even though they are means to the same end. However, while many would view use of testosterone as unethical in a sports context, they would view it as completely ethical in a medical context to treat an illness for which it was shown to be safe and effective. Both the means and the ends must be considered when evaluating the ethics of using medicine, biotechnology and nanotechnology. Challenging ethical work is involved in distinguishing between the appropriate means towards a good end, and whether a means appropriate in one context is ethical in another context.

Resolution of this issue is not just a matter of ethical concern. National health services and health insurance companies try to maintain a firm line between treatments they will contribute towards and enhancements for which patients must pay for themselves. The Normal Function Model, proposed by Norman Daniels, is frequently used to guide such determinations. According to this model, 'the central purpose of healthcare is to maintain, restore, or compensate for the restricted opportunity and loss of function caused by disease and disability' (Sabin and Daniels, 1994, p. 10). Thus, medicine should seek to help people function 'normally', defined as 'restoring an individual's functional capacity to the species-typical range for their reference class, and within that range to (the bottom of) the particular capability level which was the patient's genetic birthright' (Juengst, 1997, p. 129). Any therapy that attempts to take a person to the top of the normal range of that ability for that age or gender, or to exceed the current limits, is viewed as an enhancement, and falls beyond the proper domain of medicine or healthcare in general.

Hence, cosmetic surgery is considered an enhancement therapy in this model and is not covered by most health insurance or national health plans. Breast implant operations, facelifts and other changes in personal appearance are available to those who wish to pay out of pocket. In contrast, the same procedures might be covered by insurance plans if used to correct disfigurements caused by congenital disease or accidents, with the understanding that the procedure is restorative. In this model, procedures are viewed as 'medically necessary' treatments if they bring people's appearances closer to, or into, the range of what is viewed as normal, or if they return their appearances closer to what would have been natural for them before an accident.

The Normal Function Model has many advantages, foremost being its apparently objective nature. Medicine traditionally addresses physiological problems. If a bone is broken, medicine helps to heal the fracture and restore normal functioning. Cancer left to itself will often cause pain and shorten someone's life, so medicine tries to cure it. Medicine enters a different realm when it is asked to deal with social, psychological or spiritual problems.

When medicine is asked to address behavioural problems, to help people develop intellectually or to change their body shape to avoid social discrimination, it is being directed by values other than the promotion of physical health. Drugs and medical technology can enhance classroom behaviour, facial beauty or test scores, but these uses raise ethical concerns. One is that medicine may not be the best means to enhance these traits. Another is that this broadens the goals of medicine and biotechnology. The goal of treatment gives medicine and biotechnology a relatively clear purpose. Medicine is changed when used for other purposes: to help people look or feel better; to help evolve a new type of human; to extract confessions; to execute death-row inmates. Medicine is changed when it becomes a powerful ideological tool in the hands of those who control it or buy it, rather than a tool in the service of people's health needs. That change has important consequences.

MEDICALIZATION

The treatment-enhancement distinction is based on the assumption that medicine should be focused on treating illness or restoring health, and not on other goals related to enhancement.

Medicalization is the term used to describe the spread of medicine and biotechnology, including nanomedicine, beyond its traditional bounds to address areas of life which are not usually seen as medical issues. It is 'a process by which nonmedical problems become defined and treated as medical problems, usually in terms of illness and disorders' (Conrad, 2007, p. 4). The term 'medicalization' is usually used by authors who are critical of the process. Once a problem becomes defined in medical terms, the use of medicine, pharmaceuticals or biotechnology to 'treat' the condition appears legitimate.

Recent areas of medicalization include ADHD, mild depression, erectile dysfunction (ED) and short stature. Not long ago, if a child was disruptive in a classroom or acting inappropriately in public, people might have said to themselves, 'That child needs to learn to control himself', or possibly, 'Why don't that child's parents discipline him?' Over recent years, what was seen as a behavioural problem has been medicalized. Some children are now diagnosed with ADHD and the recommended response includes Adderall, Ritalin or other drugs. Today, when a child is seen misbehaving, some will wonder, 'Why is that boy not on medication?' or 'That girl's medication needs to be adjusted'.

In some cases, this is completely appropriate. The physiological basis for certain behaviours is understood, and to overcome properly diagnosed disorders some people need medical help. Prescription drugs have done much good for many people. Medicalization points to a broadening of the response beyond such clearly legitimate cases. The concern is that too many cases of misbehaviour will be treated by drugs or other medical therapies. At the same time, other ways of addressing the problem will be ignored, minimized or forgotten about.

The process of medicalization lays the groundwork for accepting the use of enhancement therapies. Drugs for ADHD have been shown to be effective in helping children remain attentive in the classroom. It didn't take long for people to think about using these to help normal students become more attentive.

Studies have found that many American college students use prescription stimulants to enhance their academic performance. A survey of almost 2,000 students found that 4 percent had a legitimate prescription for an ADHD medication, while 34 percent had used such medications illegally (DeSantis *et al.*, 2008). The goals were primarily to help students stay awake, concentrate better or memorize, with 7 percent looking for a high. The medications were reported as being very easy to obtain, primarily from friends, or friends of friends. One student told the researchers, 'Today it seems like everyone is ADHD, aren't they? Everybody is medicated on something' (*ibid.*, p. 320).

The respected scientific journal *Nature* carried out an informal, non-scientific survey of readers, asking them about their use of three prescription drugs for cognitive enhancement: Ritalin (for ADHD), modafinil (to overcome excessive drowsiness), or beta-blockers (heart medication also used to reduce anxiety). Twenty percent of the respondents stated that they used one or other of these medications for non-medical purposes to improve concentration, focus or memory (Maher, 2008). Many others added that they also used Adderall (mixed amphetamine salts). The informal nature of the survey means it cannot be generalized to estimate what proportion of academic scientists in general use these drugs. However, it does reveal that quite a few academics are using cognitive-enhancing drugs and obtaining them illegitimately. About 80 percent of the respondents believed adults *should* be able to take these drugs if they wanted.

Drugs being used for enhancement were developed to treat other disorders. Ritalin was developed to treat ADHD, and found to help students study; modafinil to treat narcolepsy, and found to help sleep-deprived professionals; Viagra to treat heart conditions and found to treat ED and now used to enhance sexual pleasure. Neuroscientist Anjan Chatterjee (2007) predicts that cognitive enhancing drugs, devices and procedures, including those developed through nanotechnology, will be used more widely as they become available.

The pharmaceutical industry has noticed this trend. 'This is an area where demand is going to shape the kinds of products that are made available and the kinds of world that they make' (Farah, 2008). New classes of drugs called ampakines and protein modulators are being developed 'to augment normal encoding mechanisms. They might also apply to disease states, as an afterthought' (Chatterjee, 2007, p. 130). Producers of these drugs realize there is a huge market in enhancements. Funders of nanomedicine are also likely to be attracted to enhancement therapies because of the large market and the potential for faster return on their investments. Nanomedicine is being used to produce dietary supplements (of unknown effectiveness) that claim to enhance people's lives rather than treat any illness (Erickson, 2009). As research funding, resources and creative energy are dedicated more and more to developing enhancements, less will be directed to the diseases which are killing people by the thousands every day around the world. This is an issue of global justice.

PRACTICAL STEPS TO DETERMINING IF ENHANCEMENT IS INAPPROPRIATE

A firm line between treatment and enhancement is difficult to draw, but the distinction remains useful. Many of the criticisms point to problems with finding a straightforward method of classifying each therapy. One therapy can be used as a treatment in some people or as an enhancement in others, and some treatments cause enhancement simultaneously. Rather than use the distinction to classify therapies, it will be used here to draw attention to a variety of issues concerning any therapy.

Five questions will be presented here. In answering them about a particular use of a therapy, clarity should develop regarding the degree to which that use raises ethical concerns that must be considered carefully (O'Mathúna, 2002). The goals and means involved must be examined, as must the broader consequences and implications. This approach may not give black-and-white answers in every case, but it can help clarify why a therapy is being used and reveal the extent to which that application may be ethically problematic.

1. *Is the therapy being used to treat disease or restore normal function?*

Treating illness and disease and restoring normal function are part of the ethical mandate upon which medicine and healthcare are based. If this is the goal of using a therapy, the usual ethical questions about safety, effectiveness and informed consent apply. Patients' choice and values must be involved in deciding whether a treatment is appropriate for them. Many of the applications in nanomedicine fall into this category and will not raise additional questions beyond those that apply to other drugs and medical procedures.

When a therapy seeks neither to correct biological disease nor to restore normal function, additional questions should be raised. Its use may be for enhancement which may or may not be appropriate. Vaccination is one such example. As mentioned earlier, vaccines enhance the immune system so that a future illness may be prevented. Nanoparticles are being investigated for the development of vaccines that can be applied topically or orally, thus making them more cost-effective and less painful for patients (Kubik *et al.*, 2005). Some reject the value of the treatment-enhancement distinction because it might categorize such vaccines as an inappropriate enhancement therapy. This goes against our intuition that vaccines are an ethical component of modern medicine (Juengst, 1998). The goal of vaccination is the same as that of treatment: to reduce the incidence or impact of illness by preventing disease. As discussed earlier, enhancement *per se* is not unethical. Thus, the goals of vaccination are in line with the goals of medicine, and should be seen as ethical. Decisions to use these therapies, especially the newer nanoparticle-based vaccines, should then be based on the usual medical criteria of effectiveness, safety and informed consent.

2. *Is the therapy directed towards a legitimate end, but neglects the value of non-medical means towards that end?*

Those who question the treatment-enhancement distinction note that enhancement is an accepted part of human society. Ethical concerns are not usually directed against the goal, but against the means to obtaining that goal. Physical fitness is a good goal, and some people enjoy exercising to achieve that goal. Others view exercise as 'a time-consuming burden reluctantly undertaken as a means to achieve certain ends' (Bostrom and Roache, 2007, p. 128). For these authors, medical interventions should be available to help people get fit safely and conveniently. 'Increasing one's strength by taking a drug, for example, would dispense with the need to spend hours working out at the gym or exercising with a physiotherapist, freeing up time for other activities' (*ibid.*). Bud's testosterone pump would meet this need (Stephenson, 1995).

Daniels' Normal Function Model works well when applied to traits for which objective measurements can be made. A person with an infection can be diagnosed and treatment given until the infection is gone (or under control). The levels of hormones, cholesterol, blood pressure and other physiological parameters have median (normal) values, and medical treatments attempt to bring people into these ranges (allowing for individual variation).

Some traits appear to have no higher limits. For example, 'humans can never seem to have too much intellectual, moral, or spiritual experience' (Juengst, 1998, p. 36). A therapy that improves one of these traits is not a treatment if the person already has a normal ability. What could be problematic about helping someone become even more intelligent? If a nanotech implant replaces hours in the gym, why is that a concern?

Personal character traits for which continuous improvement seems warranted are significantly different from physiological capacities like blood pressure or hormone levels. The means by which personal improvement is sought are just as important as the ends themselves. The means 'are a part of the very definition of the activity, and transforming them transforms, and can devalue, the activity itself' (Brock, 1998, p. 58).

Education as a means of enhancing memory and intellectual abilities is not preferred just because it takes longer or requires hard work. The process of education includes learning to study, exercising one's intellect, memorizing, revising for examinations, developing an appreciation for long-term investment, accepting the value of hard work, and developing other skills. All these contribute to the goal of education. An enhancement therapy that grants someone greater attention skills will not give someone everything that is involved in education. It could even divert people away from some of what is vital in the educational process. 'Technology, precisely because of its power and efficiency, seems to cheat us of the experience of accomplishment, which is something valued in distinction from the achievement of the end' (Cole-Turner, 1998, p. 155).

When people hike to the top of a mountain, they experience great joy in the vistas (O'Mathúna, 2002). Being above the clouds or seeing the valley below can be a wonderful experience. Some of the experience at the mountaintop comes from realizing that getting there was quite an accomplishment. The climb may have been tough at times, but perseverance and stamina won out in the end. The path up the mountain, with all its sweat and dust and blisters, along with glimpses that foretold of the beauty awaiting at the top, was part of the experience of looking out from the peak.

A helicopter could put someone on the top of that same mountain far more quickly and efficiently. That person could see the same views as the dirty, tired and thirsty hiker. Yet the helicopter passenger would miss out on some of the experience of hiking to the summit. The means by which people achieve the same end makes for a very different experience. The appropriateness of each means depends on the overall goal of reaching the summit.

If another hiker lies injured at the summit, and the helicopter brings a doctor to his aid, the new technology is ethically appropriate – if not mandatory – for achieving the goal of helping the injured hiker. If, on the other hand, the goal is to develop better hikers, experiencing the hike is an integral part of the process. In that case, technology that grants someone the goal without going through the process defeats the purpose of developing better hikers.

This approach also accepts that some people may not have the ability to attain certain goals. For example, someone wheelchair-bound may never be able to hike up certain mountains. Providing technology that assists them to reach the summit would be appropriate. Their experience of the mountain will be different, but this example returns us to more of a treatment scenario. The technology overcomes a limitation imposed by a disability.

Deciding about a therapy means 'we should consider carefully the value that is distinctive to the old means, value that could be lost if we simply replace old means with new' (Cole-Turner, 1998, p. 156). The old means to enhancing memory involved much time, repetition, and humility to submit to teachers and trainers. Taking a pill or inserting an implant to produce the same results would cheat a person out of the important lessons involved, and the skills needed to become an educated person.

An emphasis on medical therapies may also overlook obvious alternatives. Sleep-deprived doctors and nurses perform less effectively and make more errors. Research has examined drugs like modafinil to enhance performance. The first controlled study of a 40-minute nap during night shift found many valuable improvements in performance (Smith-Coggins *et al.*, 2006). Another study found that naps were as beneficial as modafinil, and a combination of both was most effective (Batéjat and Lagarde, 1999). In pursuing enhanced performance, however, naps during night shift were viewed as 'not practical' (Gill *et al.*, 2006, p. 158). Yet, presumably, doctors and nurses are given breaks for other necessary bodily functions.

Medical enhancement runs the risk of formulating our problems in primarily mechanistic terms. When emotional and psychological problems are viewed primarily as chemical imbalances, the role of cognitive reasons and beliefs can be neglected (Freedman, 1998). If people's poor self-image is based on mistaken beliefs about themselves, enhancing their biochemical levels will not solve the ultimate problem. A therapy that (presumably) improves intelligence or character attempts to replace educational and moral development with a quick fix. Whether or not it actually achieves the end is something that must also be considered. A medicalized enhancement could be a means that actually interferes with attaining the ultimate goal being sought. This brings us to the next question.

3. *Is the therapy actually able to deliver the promised enhancement?*

Philosophers are known for creating thought experiments and ethical dilemmas. These have value as teaching tools but have their limitations, especially if they are unrealistic. Some ethical discussions about enhancement proceed as if the therapies worked perfectly. Many do not. The discussions are still important, but we must always return to reality. Just as science fiction scenarios can have a role in thinking through ethical dilemmas, their value arises when we return to the real world.

Enhancement therapies are often presented in unrealistic ways that defy current capabilities. For example, advocates of cognitive enhancement present modafinil as a drug that 'has memory-enhancing as well as alertness-enhancing effects' (Bostrom and Roache, 2007, p. 136). Professionals working in jobs with long hours are using modafinil. However, its ability to enhance memory and alertness is not as clear-cut when the research is examined more closely – and it is not without adverse effects.

Modafinil was originally developed to treat narcolepsy, a chronic condition where people are excessively drowsy during the day and fall asleep inappropriately. It is approved as a treatment for several medical conditions involving excessive daytime sleepiness (Kumar, 2008). A number of studies have found that it improves mood, fatigue, sleepiness, and performance on various cognitive tests including those for reaction times, logical reasoning, short-term memory, and vigilance.

One study examined modafinil's effectiveness in improving the cognitive performance of sleep-deprived doctors (Gill *et al.*, 2006). The study has been widely cited in ethical debates, raising the question as to whether doctors are ethically obliged to take performance-enhancing drugs to improve their practice. Some suggest they are, though only when the benefits are substantially large (Greely *et al.*, 2008). The study involved only 25 doctors, a small number for demonstrating effectiveness. Other studies with sleep-deprived pilots have involved ten or fewer participants (Caldwell *et al.*, 2004). The doctors in the study had completed a night shift, and were finished with all patient-related activities (Gill *et al.*, 2006). They participated in interactive workshops, and afterwards took several cognitive tests. The doctors reported being better able to remain attentive during the classroom exercises, and made significantly fewer errors on some, but not all, of the tests given. The study did not involve assessment of the physicians doing any patient-related activities, which would be important to know before suggesting that these drugs help doctors perform better.

Modafinil has limitations. The study with physicians found that those taking modafinil had more difficulty getting to sleep when they later went home (*ibid.*). The study with pilots found that subsequent doses have less effect on performance, leading the researchers to worry that users might increase their doses (Caldwell *et al.*, 2004). Concerns have been expressed about whether people could become dependent on modafinil. How it works is not known, but originally it was believed not to involve dopamine, making it unlikely to be addictive. More recent studies, however, have found that dopamine may be involved somehow

(Kumar, 2008). A number of other side-effects have been reported with modafinil, including insomnia, headache, decreased appetite and allergic skin rashes (*ibid.*). In the *Nature* survey of academic scientists, about half of those using a number of cognitive enhancing pharmaceuticals reported unpleasant side-effects like headaches, jitteriness, anxiety and sleeplessness (Maher, 2008).

A general concern with drugs like modafinil is that chronic sleep deprivation is bad for us. These drugs don't prevent this, but help people feel and function better when they are sleep deprived. As one neuropsychologist stated, 'We really know very little about the short-term, and even less about the long-term, effects of these drugs in normal healthy people' (Farah, 2008). Yet she and others endorse their general use by adults (Greely *et al.*, 2008).

Beyond the direct effects of enhancement therapies, concerns arise about unexpected effects. For example, when mice were genetically enhanced to improve their cognitive abilities, they unexpectedly became more sensitive to persistent pain (Wei *et al.*, 2001). When humans use cognitive-enhancing drugs, they may change other traits that influence their work. The medications could impact creativity, flexibility or affect, and this could have important implications for overall work performance. Since improved work performance is usually given as the overall goal of these therapies, their ability to provide overall enhancement is unknown.

These findings point to the fact that current performance-enhancing agents are far from ideal. A review of the drugs, implants and procedures currently available for intervening in the brain came to a sobering conclusion (Merkel *et al.*, 2007): the neuroscientists, philosophers and legal theorists found that all current interventions have both beneficial and harmful effects. They can alter a person's psyche in unpredictable ways, changing their thinking, personality and behaviour. They concluded that cognitive enhancement therapies should not be provided by public health services because they fall outside the responsibility of healthcare professionals. They did accept that adults should be able to pay for their own cognitive enhancement therapies provided they are informed of all the risks and costs.

4. *Is the therapy being used for competitive advantage rather than for the elimination of something universally undesirable?*

When the goal of a therapy is not to treat a medical condition, careful examination may uncover the pursuit of a questionable goal. The desire for a particular enhancement often arises from a deeply felt need. Human growth hormone is a medically approved treatment for people who produce very low levels of the hormone. The injections help to prevent a number of serious conditions, and may also allow recipients to grow to a more normal height. The injections are ethically appropriate when given to correct this deficiency. A debate has developed over whether human growth hormone should be available to children who produce a normal amount of hormone but who are predicted to be relatively short (O'Mathúna, 1997). In addition, some athletes use growth hormone to increase muscle mass and strength. In all cases, nanotechnology may develop implants that replace the injections and deliver the hormone more precisely and efficiently.

The goal of using growth hormone for enhancement is either to overcome discrimination or to gain competitive advantage. Shorter-than-average people can suffer greatly from stigmatization and discrimination based on their height. The desire to avoid this pain by enhancement therapies is understandable. However, if the therapy is successful, shorter-than-average children will move closer to average height. If only the shortest children receive this therapy, another group of people will become the shortest and will likely suffer the same discrimination that those treated have avoided. The problem (discrimination) has not been solved, only shifted onto others.

Such uses of medicine are unjust. Other medical care may be available only to the wealthy, but there is an important distinction.

When a rich person pays for an expensive cancer treatment, for example, and assuming it works, those who cannot afford the treatment are not thereby at higher risk of cancer. When a rich person buys an enhancement that grants competitive advantages, those who cannot afford the enhancement are less able to access those benefits or more at risk of being discriminated against.

The physicist Freeman Dyson has expressed this concern concisely: 'As a general rule, to which there are many exceptions, science works for evil when its effect is to provide toys for the rich, and works for good when its effect is to provide necessities for the poor' (1997, p. 197). When medicine is used for competitive advantage, instead of for the prevention or elimination of disease, its limited resources become unjustly distributed.

Such types of enhancement provide what are called 'positional goods', where the benefit of a therapy depends on others not having access to it (Bostrom and Roache, 2007, p. 131). Performance-enhancing drugs in sports work this way. The drug will be of no benefit if everyone used it and ran one second faster. Bostrom and Roache admit that enhancement for positional goods is questionable. 'Generally speaking, the greater the extent to which some good is positional, the less reason there is for society to promote that good' (*ibid.*, p. 132). While they strongly advocate access to all sorts of enhancements, they seem to be aware of the practical difficulty. 'In practice, the benefits of many physical enhancements (except ones related to health and longevity) seem to have a very large positional component' (*ibid.*). Yet since they base their acceptance of enhancement therapies on people's right to choose their own lifestyle, they provide no method of discriminating between appropriate and inappropriate means.

The only way to avoid the unfair advantage of enhancement that promotes positional goods would be to make the therapy available to all. With growth hormone, everyone would be taller, and the shortest would still be discriminated against. In the case of cognitive enhancement, one group favours 'giving every exam-taker free access to cognitive enhancements' (Greely *et al.*, 2008, p. 704). While this addresses the issue of justice, 'the fundamental question is not how to ensure equal access to enhancement but whether we should aspire to it in the first place' (Sandel, 2004, p. 52). Part of the problem is that underlying the quest for enhancement are attitudes, values and dispositions that are highly questionable.

5. *Does the therapy reinforce suspicious or unjust norms and values, even if that is not its intended use?*

With the broadening of its goals, medicine is now being asked to change people's physique; remove the pain of prejudice and poor self-image; and enhance people's cognitive, emotional and physical traits – in addition to preventing and treating illness. Medical treatment that seeks to remove physical problems can be distinguished from enhancement where the 'problem' arises from cultural values and prejudices. 'Concerns with appearance, then, reflect the influences of social attitudes, values, and preferences' (Little, 1998, p. 163).

Shorter people may want to become taller because they believe that shorter people are viewed as deficient or defective, or at least not normal. This is very different from determining that a person's heart is not working normally and needs to be treated medically. Heart disease is diagnosed primarily against biological norms, whereas a determination that someone is too short is based on cultural values.

The real problem lies in the social attitudes that foster discrimination. While providing enhancement therapies might appear to alleviate such suffering, it is more likely to hide and even perpetuate the underlying problems. The trouble with well-intentioned enhancement therapies is that 'others see in them a legitimization of or pressure to meet norms' (*ibid.*, p. 172). One man who is 4 feet 4 inches tall reflected on how growth hormone therapy impacted him. 'On the one hand, I'm supposed to feel encouraged to participate fully in society, but on the other hand, I'm supposed to be encouraged not to be here!' (Sawisch, 1986, p. 48).

One result is that people may use enhancement therapies they otherwise would view as too risky. In the *Nature* survey of academic scientists, the vast majority (86 percent) agreed that children under 16 should be restricted from taking cognitive enhancing drugs (Maher, 2008). However, when asked how they would feel if other children at their own child's school were taking such drugs, one third said they would feel pressure to give cognitive enhancers to their own children. Such is the pressure that may arise as society becomes more accepting of 'enhancement medicine'.

The impact of increased enhancement goes beyond the individual. When enhancement therapy is viewed as the way to address social injustice, the victims are the ones who must undergo the inconvenience, expense and risks of addressing the injustice by using the therapies. 'Attempting to improve social status by changing the individual risks being self-defeating . . ., futile . . ., unfair . . ., or complicitous with unjust social prejudices' (Juengst, 1998, p. 42). The result is that discrimination remains and the complex conditions that give rise to the underlying attitudes are not addressed.

CONCLUSION

Active pursuit of enhancement is not inherently unethical. In contrast, the pursuit of enhancement, the pursuit of excellence, underlies much of what society values and admires in humanity. Scholars pursue wisdom, athletes give it all for victory, musicians strive for perfection, scientists search for knowledge, people seek spiritual insight, parents struggle for their children; all reflect the human desire to enhance ourselves and society. It is in our nature to want to improve.

But *how* we pursue enhancement matters deeply. The means matter. Some means distract us from the best means, while other means cannot deliver what they promise. Some deliver only by putting others down, and some promote questionable values.

Underlying medical enhancement – including nanomedicine for enhancement – is a problematic set of attitudes. An assumption exists that we should be able to change anything we want. Such enhancement projects keep alive the vision that we can be masters of our lives and our destinies. Michael J. Sandel is a political philosopher at Harvard University who claims that the enhancement project reveals 'a Promethean aspiration to remake nature, including human nature, to serve our purposes and satisfy our desires' (2004, p. 54). This aspiration just may not be possible.

One of the challenges of life is coming to accept that some things are out of our control. There is a wisdom is seeing that we cannot change everything. Some aspects of life are simply given. The frenzied pursuit of enhancement refuses to accept that we are all vulnerable and have our limitations. This denial has consequences. 'To the extent that enhancements overcome, or lead us to deny, the vulnerability of the body, they also foreclose the kinds of self-formation that our awareness of vulnerability makes possible' (McKenny, 1998, p. 235).

Part of this self-formation includes acknowledging 'the giftedness of life' (Sandel, 2007). By this Sandel means that our lives originated in a place that is not in our control. He finds the idea in religion, but also in the secular writings of Locke, Kant and Habermas. It leads to a 'habit of mind and way of being' which runs counter to that behind biomedical enhancement (*ibid.*, p. 96). This view reminds us that our talents and abilities are not of our own doing. We did not generate our genetic heritage. We were given much by those who helped us in our formative years. Our parents or primary care-givers were central, so were the friends, neighbours, relatives, teachers, coaches, doctors, nurses and myriad others who helped us become who we are today. This should lead to a sense of gratitude and humility. Our lives are not only ours to do with as we please. They are what they are in large part because of what we have been given. They are gifts. And as gifts, they bring responsibilities to use them well. For those to whom much has been given, much is expected. Using them well means using them for more than our gratification and pleasure. It includes using them for the good of others.

Life viewed as given or inherited contrasts with life viewed as earned or merited. Rather than a focus on what has been given, the focus can be on achievement and performance. Instead of humility, this leads to pride. Striving has a place, but striving based on gift is different to striving based on merit. Rather than seeing others as those who contribute to who I am, others become a source of competition. Ironically, when life is viewed as earned rather than a gift, the pressure mounts to continue to find ways to merit life. Rather than bringing relief, cosmetic surgery, as a form of enhancement, contributes to further suffering 'by continually upping the ante on what counts as an acceptable face and body' (Bordo, 1998, p. 202).

Keeping the focus on physical traits to be enhanced, we miss opportunities to reflect on the deeper things in life. Instead of looking for another enhancement, we could seize the 'opportunity to engage in a learning process about the inevitability of change, the impermanence of all things. Yes, and about our own physical vulnerability, and mortality. Others around us are dying too' (*ibid.*, p. 205). But in our pursuit of perfection, the vulnerability of the body and mind is seen 'as a condition to be eliminated by medicine rather than as an ineluctable feature of life to be addressed by moral or spiritual formation' (McKenny, 1998, p. 230).

The failure to deal with suffering and learn from it brings missed opportunities to learn empathy and compassion. Rather than sitting in the dust with our hurting friends, we focus on enhancements that have limited value in actually achieving their goals. Interestingly, many children with growth hormone deficiency and their parents observed that their very shortness left them more compassionate towards themselves and others (Sawisch, 1986). Coming to accept ourselves as we are is an important step in learning to accept others with *their* limitations and differences. In seeking to enhance ourselves we may make it harder to accept ourselves:

> Changing our nature to fit the world, rather than the other way around, is actually the deepest form of disempowerment. It distracts us from reflecting critically on the world, and deadens the impulse to social and political improvement.
>
> (Sandel, 2007, p. 97)

Constant striving for perfection can leave it harder to accept the imperfect. When technology is used to gain perfection, it seems natural to use it to eliminate imperfection. As we look less to the givenness of life, and more to merit and effort, it becomes more difficult to accept the givenness of disability and suffering. As we will see in the next chapter, the pursuit of genetic improvement and perfection has so far led only to our ability to deselect those who are obviously imperfect.

Enhancement therapies are driven by a life dominated by competition and the pursuit of perfection, not by being and participation. With that, it becomes more difficult to accept those who have less and who do not contribute. When we ask why the successful owe anything to the disadvantaged, a society built on merit has no good answer. A society built on giftedness does. Just as I did not earn my privilege, they do not deserve their lack of privilege. Just as I was given my gifts, I should be willing to give to others. 'We therefore have an obligation to share this bounty with those who, through no fault of their own, lack comparable gifts' (*ibid.*, p. 91). This mindset, therefore, directly ties into earlier discussions of global justice and leads into our final topic, posthumanism.

9 The Posthuman Future: Making Room for Human Dignity

Nanotechnology is one way by which we can alleviate our ills and improve our lot. To this extent, nanotechnology builds upon human creativity, ingenuity and compassion. These capacities express something of human dignity. Nanotechnology can be an expression of and promote human dignity by alleviating suffering and providing more dignified conditions for people to live in.

Nanotechnology in medicine has the potential to do much good by developing new and better ways to prevent, diagnose and treat illness. As examined in previous chapters, it can also generate numerous ethical questions, especially when used to enhance human beings. Many of these conundrums are not created by or unique to nanomedicine, but it has the potential to intensify and increase enhancement options.

At the extreme end of enhancement lies an even bigger ethical challenge. Some see nanotechnology as more than just a means to improve the human lot; they see it as a way to improve human beings themselves. 'Nanotechnology, in combination with biotechnology and medicine, opens perspectives for fundamentally altering and rebuilding the human body' (Grunwald, 2005, p. 197). Nanotechnology will provide implants and devices to repair, replace or enhance numerous physiological functions; sensory capabilities could be expanded (for example, by broadening the electromagnetic spectrum that enhanced eyes can 'see'); cognitive capabilities could be expanded by man–machine, brain–computer interfaces; and even completely new capabilities could be added to those possessed by human beings.

CONVERGING TECHNOLOGIES

Nanotechnology could contribute greatly to the realization of this vision of enhanced humanity. Extraordinary powers will develop from combining nanotechnology with developments in other fields, especially genetic technology. One textbook notes that 'the next fifty years will be dominated by a rapid and revolutionary advancement in both the fields of biology and nanotechnology' (Gazit, 2007, p. 123). The convergence of these fields gives nanobiotechnology. 'The ultimate goal of nanobiotechnology is the production of functional biological-relevant machines at the nano-scale' (*ibid.*, p. 121).

An even broader convergence was highlighted at a workshop sponsored by the US National Science Foundation and the Department of Defense and its subsequent report, *Converging Technologies for Improving Human Performance* (Roco and Bainbridge, 2003). Mihail Roco was one of the principal architects of the National Nanotechnology Initiative (NNI) and a major promoter of nanotechnology in many countries, earning him the title 'Prime Nanotech Emissary/Missionary to the world' (Edwards, 2006, p. 36). William S. Bainbridge (2003) is an enthusiastic supporter of enhancement, transhumanism and uploading of human personalities into computers.

'Converging technologies' refers to the synergism of nano-bio-info-cogno (NBIC): nanotechnology, biotechnology and genetic engineering, information technology and computing, and cognitive science. The above NBIC report brought a level of mainstream and scientific respectability to some of what had been regarded as fringe futuristic nanotechnology. For some, though, it 'reinforced the science fiction aura surrounding nanotech' (Edwards, 2006, p. 38). The NBIC report envisioned a 'new renaissance' through converging technologies 'based on material unity at the nanoscale and on technology integration from that scale' (Roco and Bainbridge, 2003, p. 2). Enhancement of human physical, mental and social capabilities is stressed repeatedly, to be realized by nanotechnology-based implants and nanoscale machines.

The highest priority was given to 'The Human Cognome Project' to understand and enhance the human mind (*ibid.*, p. 14). Others note that, 'Revolutions in semiconductor devices, cognitive science, bioelectronics, nanotechnology, and applied neural control technologies are facilitating breakthroughs in hybrids of humans and machine' (McGee, 2008, p. 207). Brains will be connected to machines and brains to other brains. The ultimate vision is that 'humanity would become like a single, distributed and interconnected "brain"' (Roco and Bainbridge, 2003, p. 6). Futuristic nanotechnology has become mainstream.

This mechanistic view of humanity is no accident. As nanotechnology has been inspired by nature, so mechanical imaginary has increasingly been used to describe nature. K. Eric Drexler describes biological molecules as nanomachines, stating, for example, that, 'Teams of nanomachines in nature build whales' (1986/2006, p. 162). Normal nanotechnology often uses the same language, with one textbook stating that there are 'nanomachines all around us' in every living thing, including almost all of our cells. They added that these are 'creations of the first and last nanotechnologist, Nature' (Ozin *et al.*, 2009, p. 691).

Some seek to learn from and imitate nature, as in designing nanoscale motors based on the flagellum in bacteria (*ibid.*). Others seek to use nanotechnology to improve nature. The emerging field of synthetic biology aims to build artificial forms of biological devices and even organisms (Royal Academy of Engineering, 2009). Drexler says he wants to 'use protein nanomachines to build better nanomachines' (1986/2006, p. 73). Within these visions, the human body is seen as a machine, made up of nanomachines, and similarly available for redesign and upgrades. A nanotechnology textbook, published by an academic, scientific publishing company, proclaimed:

> Many people ask themselves whether machines will ever have the ability to think. The answer is very clear and simple. There are machines that can think! The most elaborate one is called *Homo sapiens* . . . The brain is a very elaborate machine, but it is just a machine that obeys the rules of chemistry and physics. There is no reason that such a machine will not eventually be built in a laboratory or later even in a mass-production assembly line. The bionanotechnological principles presented in this book allow [us] to envision ways to make such complex machines . . .
>
> (Gazit, 2007, p. 126).

Technological reasons may not prevent us from building brains, but ethical arguments can be advanced. The fascination with building nanomachines is lending credibility to producing parts for people and ultimately upgrading us. The arguments put forward to extend nanotechnology into this enterprise need to be examined carefully and thoroughly.

HUMAN, TRANSHUMAN, POSTHUMAN

Nanotechnology is providing the tools and techniques by which humanity may be transformed. The philosophical groundwork for this move has been developed by various groups, most notable post-humanists. These promote the vision that humans are about to direct their evolution into posthumans (Bostrom, 2004). A new species is not immediately on the horizon, but serious ethical questions need to be raised about any step in that direction and the ultimate goals. The stakes are high. Some observers warn that 'anything with the power to take us "posthuman" should be watched with a beady eye; each incremental advance should be presumed dangerous until proven otherwise' (McKibben, 2003, p. 119). Francis Fukuyama (2004) has described posthumanism as the world's most dangerous idea.

Posthumanism raises questions about human dignity and its basis. At stake is whether there is such a thing as human nature and to what extent it is tied to the human body.

> Significant ethical concerns are, however, raised by the potential for using these technologies to enhance and augment human capabilities, and by the possibility that humankind, as we know it, may eventually be phased out, or become just a step in evolution.
>
> (McGee, 2008, p. 207)

Nick Bostrom is the Director of the Future of Humanity Institute at Oxford University and a prolific writer promoting posthumanism. In 1998 he co-founded the World Transhumanist Association, now known as Humanity+. The organization

> advocates the ethical use of technology to expand human capacities. We support the development of and access to new technologies that enable everyone to enjoy **better minds, better bodies** and **better lives**. In other words, we want people to be **better than well**.
>
> (Humanity+, 2009, emphasis original)

Bostrom makes the central role of human enhancement more clear in his definition of the posthuman.

> I shall define a posthuman as a being that has at least one posthuman capacity. By a posthuman capacity, I mean a general central capacity greatly exceeding the maximum attainable by any current human being without recourse to new technological means.
>
> (Bostrom, 2008, p. 108)

The three general central capacities refer to healthspan (remaining fully healthy, active and productive, mentally and physically), cognition and emotion. Bostrom accepts that these overlap in areas, and do not include every capacity that is important to humans or posthumans.

Posthumanism is group of philosophies centred around the technological enhancement of humans. Proponents vary on the positions they hold on different issues, making posthumanism a challenging view to pin down precisely. Using technology – especially nanobiotechnology – to enhance human bodies, minds and souls is central. The term 'posthumanism' is often used interchangeably with 'transhumanism'. Both ideas have their origin in enlightenment principles of progress and rational humanism (Bostrom, 2005). These include an emphasis on reason and science, a commitment to progress, and a valuing of human existence in this life. Transhumanism was first used to describe the philosophies that went beyond humanism to take humans to a radically enhanced posthuman condition. Posthumanism is the goal, and transhumanism is the philosophy to there. For our purposes, the term 'posthumanism' will be taken to include transhumanism.

The Transhumanist Declaration begins:

> Humanity will be radically changed by technology in the future. We foresee the feasibility of redesigning the human condition, including such parameters as the inevitability of ageing, limitations on human and artificial intellects, unchosen psychology, suffering, and our confinement to the planet earth.
>
> (Humanity+, 2002)

Central to the posthuman agenda is the slowing or reversing of human ageing. The result would be that 'we could remain at our fittest and healthiest indefinitely' (Bostrom and Roache, 2007, p. 124). A technical distinction can be made between 'ageing' and 'senescence'. Our usual experience is that, as time passes, our bodies get older, frailer, sicker and eventually die. We use ageing to describe this process, but posthumanists (and others) believe that ageing and senescence can be decoupled. Ageing would then refer only to the passage of time. Senescence is the process of accumulating cellular damage and other wear-and-tear, eventually leading to our deaths. We associate ageing with increased disability and reduced productivity, but these actually refer to senescence. According to posthumanists, we can age without deterioration in health, vigour or mental acuity. The result would be human life expectancy of 'around 1000 years' (*ibid.*). That is the posthuman vision.

WHAT ABOUT ALL THE PEOPLE?

Nanotechnology is one element that is seen to play a central role in realizing the posthuman vision. The connection between nanotechnology and posthumanism is not coincidental. It goes back many decades to Drexler's original interests in life-extension and space colonization. This helps explain the comment in the Transhumanist Declaration about their concern to overcome human 'confinement to the planet earth' (Humanity+, 2002). Several billionaires privately fund posthuman research projects, including private space travel programmes, radical life-extension projects, and research on artificial intelligence (Arrison, 2008). The NBIC report also envisions space conquest, with people running profit-making mines on the Moon, Mars and closer asteroids, and sending nanotech robots off to construct 'extraterrestrial bases' (Roco and Bainbridge, 2003, p. 6).

A practical concern with posthuman life-extension is how the Earth will support all the additional people. Overpopulation could result from more people living longer and potentially having more children. Bostrom and Roache (2007) claim that this will not be a problem because, as people's standard of living increases, birth rates tend to decrease. They assume that radical life-extension will lead to higher living standards. They also note that extending one's life expectancy does not necessarily increase the time period during which people can have children.

In the event that overpopulation does develop, they propose 'a policy in which those who want to avail themselves of radical life-extension would have to agree to limit the rate at which they bring new people into the world' (*ibid.*, p. 127). How this policy would be adopted and enforced raises serious practical problems. More fundamentally, it goes against the very freedom and autonomy that underpin posthumanism. In the same article they state, 'Providing they are not significantly harming others, people who live in a liberal, democratic society are free to pursue whatever lifestyle they choose' (*ibid.*, p. 125).

Eric Drexler is more realistic in his recognition of the potential problem of overpopulation, though his proposed solution is even more ambitious. In *Engines of Creation*, he claims that nanotechnology will avert any population problem because with molecular assemblers 'people will be able to make as much as they want of whatever they want' (Drexler, 1986/2006, p. 376). He claims that the economic growth of 'recent centuries shows that the rich can get richer while the poor get richer' (*ibid.*, p. 224). This highly optimistic view of human history is difficult to reconcile with the data described in Chapter 6. His optimism extends into the future also: 'When we develop pollution-free nanomachines to gather solar energy and resources, Earth will be able to support a civilization far larger and wealthier than any yet seen, yet suffer less harm than we inflict today' (*ibid.*, p. 337). The NBIC report is similarly optimistic: 'The twenty-first century could end in world peace, universal prosperity, and evolution to a higher level of compassion and accomplishment' (Roco and Bainbridge, 2003, p. 6).

At some point, Drexler acknowledges that Earth will become overpopulated and new resources will be needed. Nanotechnology will once again come to the rescue. The endless resources of outer space provide plenty of room for human expansion. Nanotechnology and advanced artificial intelligence will allow humans to travel deep into space and take advantage of its resources. Replicating assemblers and cheap spaceflight will remove all resource worries. Access to space colonization 'will burst our limits to growth, since we know of no end to the universe' (Drexler, 1986/2006, p. 339).

Space exploration leads Drexler to the need for cryonics. Freezing had long been suggested in science fiction as a way to allow people to pass the years or decades spent in space travel. Others see cryonics as a way to preserve someone until future technology is available to cure their ailments. Drexler calls the process 'biostasis' and proposes using nanotechnology to ensure people are revived safely (*ibid.*, p. 282). As the person is thawed, trillions of nanobots will be injected into the person's body to search for molecules or cells in need of repair. Cell repair nanobots will 'work under the direction of on-site nanocomputers' (*ibid.*, p. 294). Gradually the whole body will be checked and repaired, and the person awakens.

Drexler has long been involved with groups advocating cryonics and space exploration (Regis, 1990). Drexler mentions Robert Ettinger's 1962 book, *The Prospect of Immortality*, which gave rise to the cryonics movement and incorporated nanobots. 'Surgeon machines, working twenty-four hours a day for decades or even centuries will tenderly restore the frozen brains, cell by cell, even molecule by molecule' (Ettinger, 1962, quoted in *ibid.*, p. 88). Nanotechnology gave the cryonics and space travel communities the devices they needed to overcome significant limitations and challenges. As one critic put it, 'it seemed that nanotechnology could accomplish virtually all things, quickly, cheaply, and without your having to lift a finger' (*ibid.*, p. 125).

SCIENCE SERVING POSTHUMANISM

Little wonder, then, that many involved in normal nanotechnology want to dissociate their projects from those of molecular manufacturing (Smalley, 2001). However, highly influential advocates now promote a vision completely compatible with posthumanism, as evidenced by the NBIC report. Nanotechnology is being conscripted into the realization of the posthuman. Many technical developments are anticipated to be achievable within ten or 20 years. While not reaching for Bostrom's 1,000-year lifespan, advocates claim NBIC technologies will deliver active and dignified lives 'far into a person's second century' (Roco and Bainbridge, 2003, p. 19). Others, such as Cambridge University's Aubrey de Grey, claim that nanobiotechnology can stop senescence so that people remain vigorous 'indefinitely' and attain 'ages many times what we reach today' (2005, p. S51). He also believes the first person to live to be 1,000 years old may be 60 already (de Grey, 2004).

Some US federal agencies appear to have taken up such optimistic projections. The National Cancer Institute (NCI) is the US federal government's main agency for cancer research and training. The NCI has created an Alliance for Nanotechnology in Cancer to harness the power of nanotechnology 'to support the NCI Challenge Goal of eliminating suffering and death due to cancer by 2015' (NCI, 2006). Such an extremely ambitious goal appears more compatible with futuristic nanotechnology than the present-day realities of the world of cancer patients and medical researchers.

While futuristic nanotechnology and posthumanism are frequently dismissed as science fiction, their similarity with fictional scenarios is not the problem. As has been shown throughout the book, mainstream scientists sometimes use science fiction to inspire innovation, to help the public connect their developments to more familiar concepts, or to win support from funders. The fundamental problem is that these visions arise from and move towards a completely different world-view that has serious implications for humanity. They want to move humanity towards a posthuman future, achieved through: 'Evolution transcending human cell, body, and brain' (Roco and Bainbridge, 2003, p. 23). It is a big idea. 'Ideas matter, and sometimes they can be dangerous' (Jakesevic, 2004, p. 32).

AUTONOMY

A central argument in posthumanism is that people are free to choose whatever lifestyle they desire, even if others question its value. At the core of the NBIC project is, 'The right of each individual to use new knowledge and technologies in order to achieve personal goals, as well as the right to privacy and choice' (Roco and Bainbridge, 2003, p. x).

This is the same argument from autonomy we saw in the previous chapter for the pursuit of more modest enhancement. People choose lifestyles that others view as questionable or worthless, spending much of their lives doing things like watching TV, engaging in extreme sports or meditating in isolation.

> That there may be reasons to believe that an extremely long-lived life would not be worthwhile, then, does not in itself justify preventing those who wish radically to extend their lifespan from doing so, if the means of doing so and the resulting extended life do not significantly harm others.
>
> (Bostrom and Roache, 2007, p. 125)

Nonetheless, Bostrom believes 'it could be very good *for us* to become posthuman' (2008, p. 108). He claims that at least some possible posthuman modes of being would be very good for humans to pursue. While people might initially question extreme longevity, Bostrom argues that current practices reveal that many people already demonstrate that they value long life. He claims that most people already desire more of the three general capacities already mentioned that posthumanism plans to significantly enhance (lifespan, cognition and emotion).

Bostrom assumes that since many people pursue enhancement in each of these areas they would support posthumanist ideals. However, this is not borne out by the little research that has been done on the subject. A series of consultations carried out in four UK locations asked people about their attitudes towards various applications of nanotechnology in healthcare. Nanotechnology was viewed very positively for treating diseases and injuries, and replacing ailing body parts. However, people had the most negative view if nanotechnology was used for enhancement, 'transhumanism' or creating 'better humans' (Bhattachary *et al.*, 2008, p. 15). These were perceived negatively because they were viewed as based on self-interest and would likely be distributed unfairly.

Bostrom is well aware that even if people want a posthuman future, pursuing it or attaining it is not automatically ethical or even valuable. He merely wants to argue that it is 'plausible that being a posthuman could be good' (Bostrom, 2008, p. 123). He claims that only a lack of imagination could lead someone to believe that current life 'already contains all the most valuable and worthwhile modes of being' (*ibid.*, p. 122).

Bostrom's general point here cannot be denied. Based on what we know is good about life, would not more be better? If good health is valuable for 80 years, why would it not be valuable for 180 or 800? We get a taste of chocolate and decide it is good, very good. We can imagine the pleasures of a life eating lots of chocolate. It is this seeming reasonableness of posthumanism that Fukuyama finds dangerous, where 'we will nibble at biotechnology's tempting offerings without realizing that they come at a frightful moral cost' (2004, p. 42).

Eating more and more chocolate is not necessarily good for us. Life is more than pleasure, and good things often must be balanced against other good things. Our good also should be balanced against the good of others. Our globally interconnected world should make it very clear that we cannot pursue our lifestyles in isolation from the rest of the world, whether they live next to us or on the other side of the world. It is very telling that posthumanists accept that radical life-extension will be available first for developed countries where populations are in decline, and that only the rich will be able to afford such technological enhancement (Bostrom and Roache, 2007; Naam, 2008).

Posthumanism appears to accept that those who have, want more. Bostrom notes that those with high capacities typically want to excel further. Good musicians want to be better; fast runners want to run faster; great scientists want to win bigger grants or a Nobel Prize. Therefore,

> those who have a certain high capacity are generally better judges of the value of having that capacity or of a further increment of that capacity than are those who do not possess the capacity in question to the same degree.
>
> (Bostrom, 2008 p. 118)

How different this is to the giftedness idea developed at the end of the previous chapter. Do we want a society where the gifted view themselves as entitled to more; or one where they are grateful for what they have and thereby inclined to share with others? As discussed in Chapter 6, we live in a two-tiered world, where some people's life expectancy is double that of others. Posthumanism, with its attraction for nanobiotechnology, runs the danger of fusing those differences into our genes, giving a biological basis for separating us into humans and posthumans. But before looking at these implications, we first need to look at the science involved in life extension, and whether the posthuman project is practically feasible.

LIFE-EXTENSION RESEARCH

Researchers outside the posthuman community have been examining ageing and what can be done to slow senescence. S. Jay Olshansky is a respected researcher of ageing (called gerontology) who has published widely on the topic of life-extension. Olshansky and many other researchers are looking for ways to impact senescence, not just treat the illnesses that currently come with ageing.

Olshansky and 52 leading gerontologists examined the many claims being made about therapies to slow, stop or reverse ageing, and then concluded: 'The biomedical knowledge required to modify the processes of ageing that lead to age-associated pathologies confronted by geriatricians does not currently exist' (Olshansky *et al.*, 2002b, p. B293). They noted that, throughout history, people have claimed to have found the key to long life; and current claims 'that it is now possible to slow, stop, or reverse ageing through existing medical and scientific interventions . . . are as false today as they were in the past' (*ibid.*). Another group of 28 gerontologists responded to Aubrey de Grey's proposal for research to stop ageing (abbreviated as SENS) by stating,

> Creative testable ideas are the lifeblood of scientific progress. In our opinion, however, the ideas of the SENS programme in which de Grey expresses such blithe confidence are not yet sufficiently well formulated or justified to serve as a useful framework for scientific debate, *let alone* research.
>
> (Warner *et al.*, 2005, p. 1,008)

They used such strong language because they fear the public is being misled by proponents of anti-ageing medicine.

THE SCIENCE OF AGEING

Human life expectancy is the average number of years a person of a certain age can expect to live. During the twentieth century, the life expectancy for children born in the USA increased to 77 years, a remarkable increase of 30 years. The 52 gerontologists cited above noted that this increase was due primarily to a reduction in infant mortality, thus making it unlikely that such an increase will occur again for those currently alive. 'The quantum leap in life expectancy achieved over the last 100 years is an unprecedented anomaly in a human history better characterized by fluctuating, stagnating, or slowly rising trends in life expectancy' (Carnes *et al.*, 2003, p. 42).

Further increases in life expectancy will be difficult to achieve. Posthumanists acknowledge that curing all heart disease would increase life expectancy in the USA by about seven years, and curing all cancer would result in a gain of about three years (Bostrom and Roache, 2007, p. 124). 'In fact, even eliminating all ageing-related causes of death currently written on the death certificates of the elderly population will not increase human life expectancy by more than 15 years' (Olshansky *et al.*, 2002b, p. B293).

Posthumanism is based on an assumption that life expectancy has increased at a steady rate and will continue to do so. This leads Bostrom and Roache (2007) to suggest that life expectancy in the USA will be 100 years by around 2060. They believe ageing will be slowed and eventually will be practically stopped, allowing humans to reach a life expectancy of around 1,000 years.

Gerontologists disagree, pointing to research on several species showing 'that the bodies of living things (including humans) are subject to biological warranty periods that limit the duration of their lives' (Carnes *et al.*, 2003, p. 40). Data from numerous species show a remarkable similarity in how life expectancy is limited by the age at which reproduction starts and stops (puberty to menopause). The increased incidence of various diseases as organisms age can be attributed to accumulated damage at the molecular, cellular and organ level.

Posthumanists do not claim that life expectancy currently can be increased. They want society to invest in developing the means to attain radical life-extension. They support projects like those described in the NBIC report and de Grey's SENS. In evaluating projects, both the ethical and practical dimensions must be examined. Part of the ethical evaluation includes determining how realistic it is that the goal will be attained. This is especially important when it comes to allocating scarce research funding and resources. If research is unlikely to produce benefits in the near future, or if the goal is unattainable, funding it is unjust when other research is likely to produce benefits for pressing issues – such as the survival of those who are already alive. It is therefore ethically significant in evaluating posthumanist claims that a large amount of data contradicts the whole enterprise. Instead, the view that human lives have an upper age limit is in fact 'a universal and undeniable biological reality – indefinite survival is not possible, and the duration of life would remain limited by biological constraints even if every cause of premature death could be eliminated' (*ibid.*, p. 43).

REDESIGNING HUMANS

The quest to improve the design of human beings underlies much of the enhancement and posthuman agenda. A widely publicised exercise was carried out to determine what humans might look like if re-engineered to remove 'design flaws' that affect health and ageing (Olshansky *et al.*, 2003). Their purpose was not to design the perfect human, as the authors acknowledged that many alternative models could be proposed. The purpose was to demonstrate how the existing human body is not well designed for living much beyond current life expectancy. They also wanted to critique the claim that ageing is 'a disease that can be reversed or eliminated' (*ibid.*, p. 95). The design flaws cause our bodies to wear out, become diseased and die, no matter what sort of life-extension therapies are available.

The authors revisited this project in 2007 and added an important point to each of their design fixes. Rather than just present the flaw and its correction, they noted that every fix has a trade-off. For example, as people get older, the muscle that closes off the windpipe to food loses tone, making the seal less tight and increasing the risk of choking. The proposed correction involved raising the windpipe so it opens above the level at which food and fluids pass down the throat. Horses have this feature, which allows them to drink and breathe at the same time. However, that adjustment has a major trade-off: we would no longer be able to speak.

Every single 'correction' proposed by the authors comes with a serious trade-off. Although they did not state this in their articles, Olshansky (2008) discussed in a presentation how they had a biomechanical engineer evaluate the stability of their designs. The report found that if the redesigned humans were to hold any weight in their hands, they would topple over.

The human body certainly has anomalous features and parts that are difficult to understand. We do not know why each structure and organ is the way it is, or why each molecular mechanism works how it does. These re-engineering exercises were designed to 'spark an idea, trigger a thought, and inspire others to think outside the box' (Olshansky *et al.*, 2007, p. 34). They set out to demonstrate that the human body is not designed to last forever.

They stumbled upon how remarkable it is that the human body functions well at all. The human body is a finely tuned marvel of biochemical engineering. Numerous functions are intricately balanced, at both nanoscale and macroscale. The implications of these findings are highly significant for posthuman enhancement projects and other genetic modifications. The complexity and intricacy of the human body do not permit much change without serious negative consequences. We tinker with ourselves with nanotools at our own peril.

CONTROLLING OUR EVOLUTION

The terminology and imagery of evolution are regularly raised in discussions of radical human enhancement and posthumanism. Drexler claims that, 'The deep-rooted principles of evolutionary change will shape the development of nanotechnology, even as the distinction between hardware and life begins to blur' (1986/2006, p. 119). The NBIC convergence is said be 'a turning point in the evolution of human society' (Roco and Bainbridge, 2003, p. x). Posthumanists look at how biological evolution describes a journey from primitive and unconscious life, to modern human life, with mind, consciousness, language and reason. Since this has been unguided, they claim that there is no guarantee that future human evolution will occur in a desirable direction. If even extinction is a possibility, '*then* the only way we could avoid long-term existential disaster is by taking control of our own evolution' (Bostrom, 2004, p. 341).

The analogy with evolution is powerful. The 'progress' from prebiotic soup to current humanity suggests that continuing on the same road will lead to further remarkable progress. Arguments from analogy are extremely powerful because they compare an issue 'to something the audience is very familiar with or has very positive feelings about' (Walton, 1989, p. 256). Evolution is so highly regarded by many, especially scientists, that anything associating itself with evolution is likely to be viewed positively. This shifts the burden of proof onto anyone wanting to argue against the posthuman proposition. However, analogies break down if significant differences exist between the two concepts being compared. Such is the case with posthuman appeals to evolution.

Evolution via natural selection is based on several key principles. One is that humans have evolved over many millennia, yet posthumanism suggests that future evolution can be directed at a vastly increased pace. Another principle is that changes and adaptations occur randomly, yet posthumanism plans to design the changes. A third is that survival is determined by fitness through the process of natural selection. Those changes which allow individuals to adapt better lead to them producing more offspring. As a result, those changes become more prevalent in the genetic makeup of subsequent generations.

What biological evolution does not permit is a guiding hand designing changes directed towards a goal. Posthumanists argue that they can guide nanotechnology, genetic engineering and other NBIC technologies in redesigning humans towards the goal of posthuman bliss and immortality. They argue against leaving evolution to chance and natural selection. The analogy with evolution is therefore fallacious and a rhetorical device to hide the degree of control which they seek to have over future change.

Central to posthumanism is an assumption that evolution can be guided towards a goal. Yet central to biological evolution is the belief that evolution is not goal-oriented. Olshansky and colleagues, who make no claim to be posthumanists, admit that their exercise in redesigning humans was not based on the principles of biological evolution. They wanted to 'go beyond usual scientific reasoning and imagine how the human body might have been designed differently *if biology were goal-oriented*' (Olshansky *et al.*, 2007, p. 34, emphasis added). But it is clearly not goal-oriented, and this undermines the posthuman argument.

Posthumanists assume that humans can guide their future evolution, and by their own admission, *they* want to be the ones in control. But biological evolution makes changes naturally, randomly. Posthumanists want to make those decisions deliberately. However, they provide no evidence to show that they have the wisdom, knowledge or capability to make those decisions properly. If humanity is somehow supposed to possess this general wisdom, they admit that our record is not very good. 'When we manipulate complex evolved systems, which are poorly understood, our interventions often fail or backfire' (Bostrom and Sandberg, 2009, p. 375). Nonetheless, they want to have a go at it – practising on humans.

What posthumans are actually calling for is not naturalistic evolution, but a form of human creation. They have a goal towards which they want to direct humanity. They claim to have the wisdom to know that this is a good and appropriate goal; they claim to have the knowledge and capacity to design and implement the changes. Their argument is more in keeping with a creation narrative than with an evolutionary one. The biblical account claims that God had a goal in mind, making humans according to his own image. He took the natural resources available and fashioned them into Man. He then took the man and made some modifications to give Woman. The posthumanists would have us believe that they can start with the raw material of existing human beings and fashion them into posthumans, made in the image of themselves.

Such activities put humans into the role most cultures have attributed to God: that of creating and fashioning humans according to his image. Posthuman evolution is humans playing at being creators. The notion of playing God is not a call to avoid interfering in nature. Every religion has God calling on humans to interfere in life to do good for their fellow humans.

Playing God is a way of expressing that humans are attempting to usurp a role that they have neither the capacity nor the knowledge to perform wisely. This theme is explored in Mary Shelley's 1818 classic *Frankenstein*, with disastrous consequences (2003). We have developed great wisdom and technology for treating illnesses and protecting one another from disasters. We have also used some of those same technologies to inflict pain and suffering, and cause disasters. Humans act appropriately when they take on tasks that help others and the environment to flourish and thrive. But we play God when we take on the task of reshaping human nature. And we do it badly because we are not God, often causing more pain, as Victor Frankenstein did in creating his monster.

The human re-engineering project noted that 'meddling' with the human body 'is an inherently tricky business' (Olshansky *et al.*, 2007, p. 34). The authors estimated that the only way to eliminate age-related diseases and disorders would be to have a genetic system that would be error-free. 'Such perfection would also wipe out those subtle changes and mistakes that made us what we are' (*ibid.*). Attempting to go there is tempting fate, with potentially disastrous consequences.

GENETICS AND NANOTECHNOLOGY

If posthumans are to develop, the enhancements introduced in one generation will need to be passed on to future generations. Replacing various body parts with nanoengineered parts may enhance an individual, but unless these changes impact the genetic material of the person's sperm or eggs, the enhancements will die with that person. The types of changes which would lead to the age extension envisioned by posthumanism require genetic modification.

> Developing a perfect human being, built to live a life that is longer and healthier than what is experienced now, would simply require that all processes designed to maintain, turn over, or repair proteins, lipids, carbohydrates, or nucleic acids be carried out with near-perfect fidelity. With this ability in place, age changes and the consequent vulnerability to age-related pathology would decrease to the vanishing point.
>
> (Olshansky *et al.*, 2007, p. 35)

Genetic engineering has allowed biologists to extend the lives of several species, including yeast, roundworms, fruit flies and mice (Olshansky *et al.*, 2002b). While individual organisms lived longer, questions remain as to whether or not the actual ageing process was changed. Much uncertainty remains about the specific causes of senescence and how to modify it. As a result, some researchers believe that ageing is such a complex, multifaceted process that we may never figure out how to safely influence it (*ibid.*). What is generally accepted is that the underlying basis of ageing is tied into our genes (Hayflick, 2007). Modifying it will require gene therapy and, if the changes are to be inherited, germ-line gene therapy.

Gene therapy as currently practised is called somatic gene therapy – it is restricted to cells that will not be passed on to offspring. In contrast, germ-line gene therapy involves genetic changes that are inherited. Such therapy has been broadly banned because of the risk of passing on unknown problems to future generations. Germ-line therapy will be required if posthumanists are to bring about the evolution of the human species. Posthumanists state that 'part of the intrigue with biotechnology' is that it will allow us to

> actually technically evolve . . . We have the technology now to alter the germ line. Somebody's going to do it somewhere. It's more a matter of figuring out how to do it safely and manage it . . . we're going to be manipulating evolution. It's not even a question of *if* anymore; it's a question of *when* and *how*.
>
> (Sirius and Duncan, 2008, p. 13)

Both the goals of posthumanism and the means to those goals must be evaluated ethically.

Many diseases arise, at least in part, because of the presence of a defective gene. Gene therapy involves various ways of getting a functional copy of the gene into the person's cells. Gene therapy has had controversial beginnings. Progress has been made, especially with hereditary diseases such as muscular dystrophy, cystic fibrosis and haemophilia. However, significant set-backs have also occurred. The 1999 death of Jesse Gelsinger, an 18-year-old healthy man who volunteered for a gene therapy study, led to a major reassessment of research protocols and oversight (Wilson, 2009). In spite of this, a young woman with painful, but not life-threatening, arthritis died during gene therapy research (Kaiser, 2007). Uncertainty continues about the precise cause of both deaths. Any medical research can be risky, but gene therapy has been particularly controversial, both scientifically and ethically.

Part of the problem, which nanomedicine is seeking to overcome, is that gene therapy uses viruses to carry genes into patients. Researchers can produce functional genes in laboratories which they then insert into viruses. When viruses called retroviruses infect a cell, they insert their genetic material into the cell's DNA. The cells then read the genes from the virus as if they were their own. When this includes the functional human gene, the disease can be completely eliminated. This is the great benefit and promise of gene therapy. When it works, the person is cured.

The first successful gene therapy was carried out on a genetic disease called severe combined immunodeficiency (SCID). The immune system in people with this rare disease does not work, and they are extremely vulnerable to infections. They live in isolation chambers, leading to the condition being called 'bubble boy syndrome'. People with SCID are missing a gene, which can be replaced using gene therapy. In 2000, the first successes were announced, and young patients were able to leave their 'bubbles' for the first time. However, while at least 17 patients now have a functional immune system, five of them have developed leukaemia. The cancer is caused when the normal gene from the therapy is inserted into the patient's DNA near the site of an oncogene. This turns on the oncogene, causing the cell to become cancerous (Staal *et al.*, 2008). Work is ongoing to figure out how to prevent this serious problem.

Nanomedicine is expected to lead to significant changes in this area by avoiding the use of viruses in gene therapy. Various nanoparticles have been tested in tissues and animals as carriers of genes. They have been successful in transferring genes into eyes, liver, intestines, muscle and other tissues. However, the research is at a very early stage. One study examined three types of nanoparticles as gene therapy carriers for eye diseases (Prow *et al.*, 2008). Chitosan nanoparticles are made from chitin, the natural polymer found in shells and exoskeletons. Based on lab studies, it was expected to be a very useful nanomaterial for gene therapy but when tested in animals' eyes, serious inflammation occurred. Nanoparticles made from a biodegradable polymer were found to be non-toxic, but did not deliver the genes into the eye cells effectively. In contrast, magnetic nanoparticles were found to be nontoxic and efficient in transferring genes to the eye cells. Such studies show that certain nanoparticles have potential in gene therapy, but much work remains to be done.

Nanotechnology is also likely to contribute to gene therapy in other ways. Implantable devices, or nanochips, were discussed in Chapter 7 for their potential in delivering drugs in controlled or varied doses (Bawarski *et al.*, 2008). These might be adapted to release genes which would effectively turn on and off other bodily functions. Such uses could range from controlling insulin production in diabetes to varying growth hormone production in athletes.

Gene therapy devices are also a part of futuristic nanotechnology. Robert Freitas (2007) has proposed what he calls a 'chromallocyte'. This nanobot would be designed to carry chromosomes which are defect-free copies of the patient's originals. The nanobots would be injected into the bloodstream, travel to the tissue or organ where the genetic defect was having an impact, enter each cell and replace the defective chromosome with a normal one. Freitas estimates that treating a large organ like the liver would require one trillion nanobots. Since they would be so small, however, this dose could be delivered in saline over seven hours. Freitas claims that his chromallocytes would be 'the ideal gene delivery vector' and 'could provide a complete and permanent cure for almost all genetic diseases' (*ibid*., p. 4).

Another area where nanotechnology is already having an impact is with genetic testing. Chapter 7 discussed 'lab-on-a-chip' devices being developed to allow many types of tests which currently require a laboratory to run. These are moving medicine closer to fast, inexpensive and accurate genetic testing. 'Simplifying genetic testing, and reducing the costs of such tests, could help pave the way toward routine delivery of such personalized care based on an individual's genetic profile' (AZoNano, 2008). *Forbes* magazine predicts that personalized genetic testing will be one of the major nanotechnology breakthroughs in the coming years (Wolfe, 2009). With that will come all the ethical issues associated with genetic testing and genetic enhancement and move us into an era like that portrayed in *Gattaca* (1997).

GENETIC SELECTION

Although nanomedicine is not necessarily directed towards genetic testing and genetic enhancement, it will contribute greatly to the tools which will make these more feasible. Many science fiction scenarios portray this type of future. Foremost amongst these is *Gattaca*, a film which 'towers above most of its science fiction contemporaries' in considering bioethical issues (Gavaghan, 2009, p. 75). The story has many themes, but prominent among them is the incorporation of eugenics into society.

Francis Galton coined the term eugenics in 1883, and later defined it as 'the study of agencies under social control that may improve or impair the racial qualities of future generations either physically or mentally' (1908, quoted in Carlson, 2001, p. 234). At the time, developed countries held to a profound belief in progress and evolution was viewed by some as a means to further human progress. For example, the poster from the Second International Eugenics Conference in 1921 had the caption, 'Eugenics is the self-direction of human evolution' (quoted in Carlson, 2001, p. 279).

Eugenic practices included encouraging those who were 'more fit' to reproduce, and discouraging those who were 'less fit' from having children. These have sometimes been distinguished as 'positive eugenics' and 'negative eugenics', respectively. They led to various practices in European and American countries, ranging from competitions to find the 'fittest families', to laws requiring sterilization of people with mental illnesses. Eugenics became entangled with ideas such as racial hygiene, anti-Semitism and Nazism to lead to wholesale slaughter of those deemed unfit, and the notion has since become viewed as inherently unethical (O'Mathúna, 2006). *Brave New World* is a novel about a society in which eugenics is not enforced at gunpoint, but is no less coercive (Huxley, 1932; 1998). An authoritarian State controls human life through social conditioning programmes and separation of classes of humans. The term has become a synonym for a society that accepts eugenics by those who view such practices as unethical.

Recently, efforts have been made to reclaim the term eugenics from the horrors of Nazi eugenics to describe what are viewed as ethical means to improve the human gene pool. A group of four internationally renowned bioethicists defended a form of eugenics that is ethical, just and 'perhaps even morally required' if its motivation is 'to endow future generations with genes that might enable their lives to go better' (Buchanan *et al.*, 2000, p. 60). This would include preventing certain births to ensure those born have 'a worthwhile life' as opposed to 'a life not worth living' (*ibid.*, p. 224). They define the latter as 'a life that, from the perspective of the person whose life it is, is so burdensome and/or without compensating benefits as to make death preferable' (*ibid.*).

The term 'liberal eugenics' has been coined for approaches where 'prospective parents should be empowered to use available technologies to choose some of their children's characteristics' (Agar, 2004, p. 2). The role of the State would be restricted to developing 'a wide range of technologies of enhancement' and ensuring citizens were informed about these in making their eugenic decisions. 'Parents' particular conceptions of the good life would guide them in their selection of enhancements for their children' (*ibid.*, p. 5). Although Agar is critical of posthumanism, liberal eugenics is compatible with posthuman goals, if not essential to their fulfilment.

This is the type of society which *Gattaca* explores successfully, in part because the film is made to look and feel much like contemporary society. The lack of restrictions on access to genetic technologies resembles the 'reproductive-genetic supermarket' in contemporary developed countries (Shapshay, 2009b, p. 87). The film follows Vincent, an 'in-valid', because he is conceived during intercourse. He is diagnosed at birth to be at risk of several diseases, with a life expectancy of 30.2 years. For their second child, Anton, the parents use *in vitro* fertilization, pre-implantation genetic diagnosis and embryo selection. He is therefore a 'valid'.

The film exposes how supposedly uncoercive liberal eugenics is 'effectively compulsory' (Gavaghan, 2009, p. 65). While Anton's parents want to 'leave a few things to chance', the clinician warns them not to give their child 'any additional burdens'. Later, we see that Vincent is prevented from attending a school or becoming an astronaut because he is an 'in-valid'. Social coercion can push genetic technologies on people, just as we noted contemporary parents feeling pressure to use cognitive enhancements on their children (Maher, 2008).

Gattaca points out the irony of how removing disease burdens and creating enhanced offspring brings its own burdens. Vincent later meets Jerome, a 'valid', who is confined to a wheelchair. Jerome, we later learn, 'suffered under a different burden: the burden of perfection'. That burden led him to attempt suicide after winning a silver medal instead of the gold he should have won. Perfect genes did not improve his self-image.

Vincent assumes Jerome's identity and shines as a trainee astronaut. In so doing, the film points out the central flaw in genetic determinism and the posthuman emphasis on enhancement. Vincent overcomes his 'in-valid' genetic heritage. Not only does he become an astronaut, he finally beats his brother Anton in a swimming race, primarily through sheer determination. As the film's subtitle states, 'There is no gene for the human spirit'. Yet even Vincent is not master of his own destiny. The authorities figure out an 'in-valid' is in the training programme and are closing in. Vincent must give a random urine sample which will reveal his true identity. Yet the doctor running the tests changes the result. Vincent's dream comes true and he goes into space. The film reminds us 'about how genes, luck and free will interact to shape our lives' (Gavaghan, 2009, p. 84). In the end, Vincent achieved his dream through an unmerited gift.

EVOLVING HUMANS

Even if the posthuman project was feasible, there are good reasons to believe it will be problematic. Posthumanism claims that evolution left on its own will eventually lead to human extinction. Bostrom calls this 'freewheeling' evolution, suggesting a car spinning out of control towards a disaster (2004, p. 341). Posthumanists plan to grab the wheel and put us back in the right direction. But biological evolution does not have a steering wheel or a destination. If posthumanists are planning to control evolution, it suggests they will select the traits of those who will evolve.

Vincent and Jerome live out the uncertainties of genetic selection in *Gattaca*. Apart from this difficulty, genetic selection neglects the value of diversity. People who are less well-endowed with certain gifts and talents should be respected and honoured for who they are. To exclude some people just because they differ from 'normal' is unethical, but so too is the desire to eliminate those whose abilities fall below some arbitrary standard. A campaign to eliminate a certain gene that is motivated by the desire to reduce diversity in people's capabilities would deprive society of the gifts given us through those who differ from us. Assuming it was technically possible, it would deprive us of important opportunities to learn to accept and encourage those who differ from us.

Posthumanists emphasise the liberal rights of individuals to choose their lifestyles and enhance themselves and their children as they wish. Yet Bostrom gives this an authoritarian dimension, in spite of the apparent contradiction. He states that for humans to take control of their own evolution 'would require the development of a "singleton", a world order in which at the highest level of organization there is only one independent decision-making power (which may be, but need not be, a world government)' (Bostrom, 2004, pp. 341–2). This global authority would be agreed to by everyone and would have no external competitors, and no internal challenges to its constitution would be permitted.

This authority is likely to come from those who have greater capacities. Bostrom claims that those who already have a higher capability in a trait are better judges of the value of that capacity and of the importance of enhancing it. This builds a degree of elitism into the programme. If posthumanism achieves its goal of starting human enhancement, the more enhanced will promote further enhancement. Rather than promoting a more egalitarian system where those who have more help to bring those with less up to their level, those with more will seek more for themselves.

It is no coincidence that many science fiction accounts of societies where humans have enhanced themselves portray situations of conflict between the enhanced and the unenhanced. Beginning with H. G. Wells and his 1898 *Time Machine*, selective breeding led to the Morlocks dominating and hunting the Eloi. *Brave New World* has the Alphas ruling over the Betas and Deltas, with World Controller Mond resembling Bostrom's singleton (Huxley, 1932; 1998). *Gattaca* uses genetic technology to separate the 'valids' from the 'in-valids'. In *Chasm City*, those with nanobot enhancements live with ease in the Canopy while the unenhanced live in the grime of the Mulch (Reynolds, 2001). Posthumanists say it won't happen again. Science fiction generally claims it inevitably will. Looking at human history, we see slavery, racism, sexism and abortion, where some humans dominate or kill those of the same species. Even today, the way developed countries ignore the plight of developing countries reflects a willingness to allow other humans to suffer while we attain our desires. Science fiction reflects insight into human nature that posthuman ideology ignores – at its peril and at ours.

Posthumanism is scientifically implausible based on what we know about the science of ageing and the biological limits of life. It is also ethically unjustifiable because it will require genetic modification of future generations before we can have any idea of the consequences for those children. The evidence we have from current experiments is that such genetic modifications are more likely to cause problems than benefits. The use of reproductive and genetic technologies shows that some are accepting aspects of liberal eugenics. A film like *Gattaca* can do much to remind us of the flaws in genetic determinism and posthumanism. Science fiction also reminds us that attempting to evolve ourselves is highly dangerous and, if it works at all, is likely to provide a biological basis for discrimination, not justice.

10 Technology and the Future: Revisiting Daedalus and Icarus

Nanotechnology has the potential for great good and great harm. The dilemma this creates is not unique to nanotechnology, but a theme running throughout the history of technology and science. Nanotechnology is the latest stage upon which the drama of human ethics is being acted out. 'Normal nanotechnology' did not seek this spotlight. It would prefer to pursue its inventions and discoveries in the quiet of the laboratory. 'Futuristic nanotechnology' will ensure this will not happen as it pulls the whole field into the limelight. That may not be fair, but it goes with the privilege of engaging with such exciting technologies and developments.

Nanotechnology is powerful and will lead to great things. It is already producing useful products with innovative applications. Nanomedicine is starting to provide new drugs and devices that will benefit many people. This work can be celebrated. But throughout history, people have recognized that humanity's next big step can be into glory or over the precipice. Literature and the arts have served an important function in addressing these concerns and raising balancing points for those driven to discover and invent. Science fiction can play such a role for nanotechnology.

Science fiction authors and scientists have dreamed the same dream. They see the good that technology can bring. Nanotechnology focuses on bringing the technology to life, while science fiction focuses on bringing life to technology – on balancing the technical dimensions with the human dimensions. Science sometimes forgets the way human values are woven into technology. It sometimes neglects the broader impact of the technology beyond the devices and the problems they solve. Nanoethics and literature are important balancing forces against the pull of technological determinism. The very successes of nanotechnology will lend credence to notions that nanotechnology can solve all our problems. We have seen how posthumanism wholeheartedly runs in that direction, but others within nanotechnology are being drawn there already. Nanoethics and literature play an important role in reminding nanotechnology to consider the human heart, to remember that while we have great powers, we sometimes lack insight.

DAEDALUS AND ICARUS

The myth of Daedalus and Icarus has served well as a cautionary tale for artisans, technologists and scientists. It warrants retelling. Daedalus was a skilled architect, sculptor and inventor – a scientist of his day. Enraged by the talent of his nephew and apprentice, he murdered him, resulting in his banishment to Crete. He ran into trouble there also, ending up imprisoned with his son Icarus. They plan their escape by making wings to fly from the island. Before setting off, Daedalus warns Icarus not to fly too low in case moisture from the sea would make his wings too heavy, and not to fly too high or the sun would melt the wax holding his wings' feathers in place.

They take off, but over the sea Icarus gets braver and exults in his new-found freedom. As put by the Roman poet Ovid, 'the boy began to delight in his daring flight, and abandoning his guide, drawn by desire for the heavens, soared higher'. Icarus soars higher and higher, the wax melts, his wings fall apart and he plunges into the ocean. Daedalus finds his son's body to bury him on shore, but curses his inventions.

Another part of the myth has Daedalus build a device that allows the Queen of Crete to copulate with a bull, giving birth to the man-bull Minotaur. Daedalus has come to exemplify the potential within human beings to use science wrongly. The Queen pushed Daedalus into satisfying her desires, and scientists today are pressured by many forces. Daedalus acceded and the results were monstrous. He then had to build a labyrinth to confine the Minotaur, who required regular human sacrifices. Tacking on fixes for past mistakes never works as well as precaution beforehand. Before taking to the skies, Daedalus warned Icarus of the dangers, yet to no avail. Once the technology was out there, it could no longer be controlled, even with all its dangers.

The story has become a metaphor in classical and popular culture for human arrogance and scientific hubris, especially in the context of new technology. Daedalus helped open the Pandora's box of interspecies breeding, and then struggled to close it. Icarus could have taken the middle route, carefully balanced between the extremes. The technology had much potential for good, but the human heart led to it causing harm.

At the beginning of the twentieth century, *Daedalus* and *Icarus* were the titles given to a debate between two of Britain's great intellectuals: the biochemist and geneticist J. B. S. Haldane (1923) and the philosopher Bertrand Russell (1924). They examined whether scientific progress would ultimately improve human well-being. Haldane was sure that science would bring increased powers, yet he was unsure whether that would improve the world – especially following on from the destructive power he saw in World War I. Haldane believed science must march on, and morality must be flexible in accommodating to its developments.

> Whether in the end man will survive his ascensions of power we cannot tell. But the problem is no new one. It is the old paradox of freedom re-enacted with mankind for actor and the earth for stage . . . [The future] is only hopeful if mankind can adjust its morality to its powers.
>
> (Haldane, 1923)

Russell disagreed with Haldane's relativistic view of morality, but was also pessimistic about human morality. Prescient of current enhancement debates, Russell believed science would produce a therapy that could make humans more generous. He didn't think we would use it, though, because 'only kindliness can save the world, and even if we knew how to produce kindliness we should not do so unless we were already kindly' (Russell, 1924). Instead, human nature will lead to the therapies being used for fame and fortune, and to enhance the ferocity of soldiers. Russell's conclusion is pessimistic.

> Men sometimes speak as though the progress of science must necessarily be a boon to mankind, but that, I fear, is one of the comfortable nineteenth-century delusions which our more disillusioned age must discard . . . science threatens to cause the destruction of our civilization.
>
> (*ibid.*)

A recent commentator on the debate notes that morality loses on both accounts. 'Russell disarms virtue, Haldane relativizes it' (Rubin, 2005, p. 90). Rubin claims their debate has set the pattern for moral evaluation of science ever since.

> Start with a little history to produce an attitude of pride that we know so much more than we once did. Look at what we know now, and stress the dangers of our remaining ignorance. Anticipate the future, and how humbled we are that those who follow us will know far more than we do if only we stick with it.
>
> (*ibid.*, pp. 90–1)

There is a power within science to intoxicate. The awe of discovery; the thrill of new knowledge; the satisfaction of success; the power of producing what was planned. These are not unique to science, but these human influences on science are sometimes forgotten. We get carried away with our successes. As one science fiction author put it, 'we tend to forget that yesterday's passionately held "solutions" often become tomorrow's problems' (Brin,

1990, p. 589). Literature can bring this dimension to the surface and facilitate ethical reflection.

Many stories do this, but Nathaniel Hawthorne's short story 'The Birth-Mark' is particularly tragic. Aylmer is a great scientist, happily married to beautiful Georgiana. She has a birthmark on her cheek which he comes to see as her only blemish and 'the visible mark of earthly imperfection' (1843/2003, p. 7). His growing dissatisfaction with the blemish develops into an obsession to remove it. Georgiana eventually accedes to his wishes and drinks the potion he has prepared. The birth-mark fades and her life fades with it. With her dying breath she possibly reminds Aylmer of Icarus. 'You have aimed loftily! . . . with so high and pure a feeling, you have rejected the best that earth could offer' (*ibid.*, p. 19).

Aylmer, Daedalus and many others have pushed on with their experiments without holding back to evaluate everything. There is a wisdom in humbly accepting that some things ought not to be done. The pursuit of perfection is not what science is ultimately about. As Sandel (2007) has pointed out, and Aylmer exemplifies, accepting the givenness of life is a sign of moral courage and an important balancing point against the reckless pursuit of perfection or power.

The givenness of life brings with it a way to accept what we do have with gratitude and generosity, and also provides a framework to work through the struggles of accepting what we would rather not have. The alternative is a never-ending search for perfection that leaves us dissatisfied with even the good things in life.

Science is a human enterprise and, like other human activities, can be impacted by personal ego, economic gain and the search for control and power. These temptations must be taken seriously and ethical guidelines brought to bear. By neglecting ethics, the cause of science is not served well, 'since science only matters in human terms if it truly serves our humanity. And that is by no means guaranteed' (Rubin, 2005, p. 91).

Aylmer also shows us the importance of remembering how much we don't know. As he experimented with his chemicals, both his assistant and Georgiana cautioned him. He replied with great confidence that they should just trust him. Such scientific hubris should set off warning signals. We cannot 'just trust' science. Against Haldane, we cannot just go there and adjust our morals to accommodate our new powers. There is much uncertainty within science, and we must take that into account when ethically evaluating scientific visions and proposals.

Donald Rumsfeld made an off-the-cuff remark when he was US Secretary of Defense that has been written off as gobbledygook by some, yet it contains some relevant insight (Steyn, 2003).

> [A]s we know, there are known knowns; there are things we know we know. We also know there are known unknowns; that is to say we know there are some things we do not know. But there are also unknown unknowns – the ones we don't know we don't know. And if one looks throughout the history of our country and other free countries, it is the latter category that tend to be the difficult ones.
>
> (Rumsfeld, 2002)

The terms provide guidance for decision-making in the face of uncertainty. Slovenian philosopher Slavoj Žižek (2004) adds a fourth category, 'the unknown known', for things we do know but either refuse to accept or act as if we don't know them. This points to the remarkable propensity for human self-deception. We want to believe we are basically good people, but literature and ethics show otherwise. Like science, we as individuals have great potential for good and evil. Blaise Pascal put it this way:

> What a chimera, then, is man! What a novelty! What a monster, what a chaos, what a contradiction, what a prodigy! Judge of all things, imbecile worm of the earth; depositary of truth, a sink of uncertainty and error; the pride and refuse of the universe!
>
> (1660, VII: 434)

Our paradoxical, almost schizophrenic, nature is acknowledged by great thinkers over many centuries from widely divergent backgrounds. We do well not to treat that as an unknown known.

The inclusion of literature and nanoethics within nanotechnology can help everyone remember this and that it must be taken into account in doing good science.

IMMORTALITY AND BEYOND

To the extent that posthumanists want to change human nature they appear to accept human limitations. But then they forget those limitations in assuming they can fashion us according to their image and in their power. They treat the dark side of human nature as an unknown known.

Posthumanism is particularly problematic when combined with the power of nanotechnology and NBIC. Many of their plans are scientifically unrealistic and dangerous. The genetic manipulation they will need to do will be done blind and may cause much harm. If they were to succeed, the likelihood is we would have another basis for conflict. The underlying problem is that no amount of nanotechnological or genetic manipulation will enhance the human heart and spirit.

The fervour for change that motivates posthumanism, and some within futuristic nanotechnology, seems closer to religious zeal than scientific dedication. Bostrom notes that, 'Many people who hold religious beliefs are already accustomed to the prospect of an extremely radical transformation into a kind of posthuman being, which is expected to take place after the termination of their current physical incarnation' (2008, p. 126). He does not advocate religion, but speaks a similar language. What religion offers through faith in the divine, posthumanism offers through faith in technology: a life of satisfaction on earth, followed by life after death.

Drexler and the posthumanists are also united in their resistance to death. They see in scientific technology the power to overcome the ultimate slight on human grandeur. Brake and Hook trace in science fiction over the last few centuries various demotions that have been brought upon humanity and our meaning in life. Earth is no longer at the centre of the Universe, nor is the Sun. Evolution has left man with no special place 'and vanishingly little evidence of a divine image' (Brake and Hook, 2008, p. 215). With the unravelling of the human genome, they predict even more turmoil.

Posthumanism offers a way back for humanity to find meaning: take control of the whole process. Believing that we got here randomly, they reject any sense of giftedness. They pursue perfection and control to the point of risking our very inheritance. One huge obstacle remains: death. The longing to overcome death and the difficulty in accepting mortality is a frequent theme in literature.

The climactic scene in *Blade Runner* (1982) provides a narrative for posthumanists to reflect on. Roy Batty is a replicant, a type of robot, who has returned illegally to Earth in search of his maker, Dr Tyrell. He, like the posthumanists, wants his 'death programme' turned off. Tyrell tells him it can't be done. They had run several experiments, all with disastrous consequences. 'To make an alteration in the evolvement of an organic life system is fatal. The coding sequence cannot be revised once it's been established,' Tyrell tells Roy. He goes on to remind Roy of his great giftedness. 'You were made as well as we could make you . . . The light that burns twice as bright burns half as long. And you have burned so very, very brightly.' Roy cannot accept this, and kills his maker.

CONCLUSION

Our discussions have moved a long way from testing nanoparticles for sunscreens and self-cleaning windows. Much of nanotechnology is concerned with products that will be perfectly compatible with the lives and values we treasure. Much of the work occurring within nanotechnology is no more ethically controversial than work with larger technology. But the ideology of technical determinism has a way of creeping into apparently mundane science. And posthumanism is attempting to spread its radical vision and impact nanotechnology's overall direction.

The final episode of *Battlestar Galactica* (2009) proposed another option. The last few survivors of the human race reach a planet like Earth. The natives are like primitive humans without technology or even language. Lee Adama suggests they destroy all their technology to break the cycle of violence.

> We can give them the best part of ourselves. Not the baggage; not the ships, the equipment, the technology, the weapons. If there's one thing that we should have learned it's that . . . our brains have always outraced our hearts. Our science charges ahead. Our souls lag behind. Let's start anew.

They destroy al their technology, yet in the final scene, several thousand years later, the technology is back. Everything looks like it was in the original Earth, suggesting that we never learn.

The physicist Freeman Dyson has stated, 'Progress of science is destined to bring enormous confusion and misery to mankind unless it is accompanied by progress in ethics' (1997, p. 99). Progress in ethics is the challenge facing us. Technology will not guarantee morality. Ethics, philosophy, theology and literature are better sources of guidance, but they also will not *guarantee* morality. The human heart, mind and soul are where moral progress is needed. A mind-set based on gratitude for our giftedness is a better place to start than a frenzied pursuit of perfection. Working with others and for others will also help.

Let us hope that we do not come to believe that our best option is to jettison all our technology. Nanotechnology has the potential to do great good for humanity if used in scientifically and ethically appropriate ways. Nanotechnology can bring great excitement in new discoveries and abilities to do new things. But we must proceed with precaution in our mind and humility in our heart. We must remember what we know and what we don't know, about both nature and human nature. And we should remember that the primary goal of science is the betterment of all humanity and the planet we call home.

Glossary

Antibodies are proteins found in the body which the immune system uses to identify unwanted objects. These are then targeted for destruction or elimination by other parts of the immune system.

Atomic force microscope is a type of scanning probe microscopy (see definition).

Atoms are the fundamental building blocks of matter. Each element (such as carbon, sodium or iron) consists of a distinct type of atom. Atoms combine to give molecules.

Autonomy is the ethical principle that competent humans should be given the freedom to make their own decisions, and that they can be held morally responsible for their actions.

Bioethics is a field of applied ethics focused on developments in biotechnology, biomedical sciences and healthcare.

Bottom-up approach to nanotechnology involves moving atoms, molecules and nanoparticles to build nanoscale devices.

Buckyballs are nanoparticles made of carbon atoms arranged in repeating hexagons and pentagons, giving them a structure that looks like a football. They were discovered in 1985 and called buckminsterfullerene after the architect Buckminster Fuller.

Bulk properties are those which large particles of a substance have and that can differ from the properties of nanoparticles of the same substance.

Carbon nanotubes are like a sheet of graphite rolled up into a long tube. In graphite, carbon atoms are arranged into hexagons and pentagons. Carbon nanotubes have a diameter of a few nanometres, and can be up to 1 millimetre long.

Colloid is a mixture where one substance is evenly dispersed throughout another. However, the particles are suspended in a colloid, rather than being completely dissolved as in a solution. Milk is a colloid.

Cryonics is a method of freezing people's bodies (or just their heads) so that they can later be revived when technology is available to treat their ailments. It is also discussed in science fiction as a means of allowing extended space travel.

Dendrimer is a complex nanoparticle with a core unit onto which several branching chains are attached. They have a tree-like structure with many cavities into which drugs and other molecules can be inserted and targeted for delivery to specific tissues.

Deontology is an approach to ethics that focuses on people's duties and obligations. It holds to universal ideas of good and evil that are discovered through reasoned reflection and an emphasis on respecting persons.

DNA, or deoxyribonucleic acid, is the molecule that carries the genetic information of an organism. Everyone has unique DNA, unless they are identical twins.

Drug delivery systems are methods used to improve the solubility and transport of drugs within the body so that the drugs are used more efficiently and safely.

Enhancement, broadly speaking, is any attempt to improve something or someone. Within bioethics, it usually focuses on the use of technology or pharmaceuticals to improve human capacities, especially improving them beyond current normal ranges.

Ethics is the branch of philosophy that studies questions of morality. It can involve examining the duties, consequences, virtues, moral values or other aspects of making decisions about what is right and wrong.

Eugenics is a range of practices motivated by the desire to improve the genetic traits and characteristics of future human generations. Positive eugenics refers to means of encouraging the 'fittest' people to reproduce. Negative eugenics refers to practices that prevent certain people from reproducing, either with or without their consent. Liberal eugenics is a recent development which promotes the use of genetic and reproductive technology to help people select the traits of their own children.

Fullerenes are a family of carbon compounds that include buckyballs, carbon nanotubes and graphene. Each of these comes in various sizes and shapes.

Futuristic nanotechnology, in contrast to normal nanotechnology, is a more radical view of how nanotechnology may some day allow the production of any item atom-by-atom.

Gene therapy is a developing area of medicine which attempts to insert genes into a person's cells or tissues to treat a genetic disease. The ultimate goal of gene therapy is to develop techniques that correct or replace defective genes in order to eliminate the disease completely.

Graphene is a one-atom-thick sheet of carbon atoms packed into a honeycomb lattice. Several sheets of graphene are stacked on one another to give graphite.

Graphite is one of the traditional forms of carbon, along with diamond and powdered carbon (charcoal). Other forms have been discovered with nanotechnology.

Gray goo is a hypothetical scenario where self-replicating nano-devices devour all matter on Earth leaving only a disorganized grey mess.

Informed consent in ethics is an expression of the importance placed on individuals being allowed to make their own decisions about their health and lives. People should be given enough information so they can understand the risks and benefits of options, and be allowed to make their decisions free of coercion or manipulation.

Lab-on-a-chip is a small diagnostic device that can completely process samples quickly and with small sample sizes. Smaller than a credit card, they can carry out numerous tests on a drop of blood or urine, rather than sending away larger samples to a laboratory.

Liposomes are tiny structures like microscopic bubbles. They have an inner area separated from the environment by a membrane that is two molecules thick (a bilayer). They allow water-insoluble substances to be suspended and transported in water.

Lithography is a general method of using moulds and stamps to produce images and patterns. Moulds can now be made with nanoscale patterns allowing nanolithography and soft lithography.

Materials science is a field of engineering and applied science which studies the relationships between a material's bulk properties and its underlying atomic and molecular structures.

Medicalization is a process by which problems and behaviours traditionally not viewed as medical issues become defined and treated as medical problems. Medicine, pharmaceuticals and biotechnology thereby gain credibility in the treatment of such conditions that previously would have been addressed through non-medical interventions such as education, counselling or social agencies.

Molecular manufacturing is a method of building items and devices molecule-by-molecule. Miniature factories and matter compilers have been proposed to allow such production, but they remain theoretical.

Molecules are made from the chemical combination of two or more atoms. Many molecules are smaller than nanoscale, though some biological molecules fall into the nanoscale range.

Nanites are a form of nanobot that typically are insect-like and used for destructive purposes in science fiction.

Nano is a prefix used for measurements that are one billionth of something. One nanometre (nm) is one billionth of a metre, or 1×10^{-9} metre.

Nanobots are, to date, fictional robots with dimensions at, or close to, nanoscale.

Nanomedicine is the use of nanoscale or nanostructured materials in medicine and in the development of new pharmaceuticals.

Nanotechnology is the broad field of enquiry seeking understanding and control of matter at dimensions between approximately 1 and 100 nanometres. It includes the practical applications of related knowledge and devices.

Narrative ethics is a branch of ethics that uses any form of literature or story to reflect on or teach ethics. Virtue ethics uses narrative to demonstrate the role of character in ethics.

National Nanotechnology Initiative (NNI) is an agency set up in 2000 to distribute the funds available from the US federal government for research on nanotechnology.

NBIC stands for nano-bio-info-cogno, or Nanotechnology, Biotechnology and genetic engineering, Information technology and Cognitive science. Developments in each field are contributing to the others so that clear distinctions between them are sometimes not possible. For this reason, they are also called converging technologies.

Normal nanotechnology, in contrast to futuristic nanotechnology, is an extension of other scientific, engineering and biomedical research, working at the nanoscale. It builds upon established scientific theories and practices to develop new technologies and applications.

Paramagnetism is a form of magnetism which appears only when the substance is present within an external magnetic field. The external field affords a way of switching the paramagnetic effect on and off.

Point of care technology refers to medical devices that allow samples to be taken and processed rapidly, providing results at the bedside or during an office visit. Their purpose is to replace the need to send samples to laboratories that can delay results.

Posthumanism is a group of philosophies united in their promotion of human enhancement. It holds that technology should be developed and used to enhance human bodies, minds and 'souls'. Through science and technology, posthumanism seeks to control human evolution, possibly leading to a new species, the posthuman.

Precautionary principle is a general ethical principle applied when decisions must be made while much remains unknown about potential risks. Formulated in different ways, the principle calls for various steps to be taken to ensure the safety of people and the environment are prioritized without inappropriately restricting development.

Quantum dots are nanoparticles containing a core of semiconductor material surrounded by a shell of other material. The particle is like a 10 to 50 nm cage that traps one or more electrons. Quantum dots made from the same material but of different sizes emit different colours of light.

Scanning probe microscopy is a way of imaging surfaces at an atomic and nanoscale level. The field includes scanning tunnelling microscopes and the atomic force microscope. Some instruments can also move atoms to align them precisely.

Semiconductors are materials with electrical properties in between conductors and insulators. Semiconductors are foundational to modern electronics, with most commercial ones made from silicon. Other types of semiconductor are being developed.

Senescence is the process of damage and wear-and-tear that eventually leads to our deaths. It allows the term 'ageing' to refer simply to the passage of time. Some hope that ageing will eventually occur without senescence.

Synthetic biology is a new field of inquiry using nanotechnology and other technologies to develop artificial forms of biological systems and organisms.

Technological determinism is a belief that technology shapes cultural values and that society's problems will be solved in the best way by technological developments.

Top-down approach to nanotechnology involves taking large pieces of materials and grinding, cutting or moulding them until their size is in the nanoscale range.

Transhumanism was used to describe the philosophies that supported belief in progress towards a future posthuman condition. Transhumanism has largely been engulfed within the term 'posthumanism'.

Utilitarianism is an approach to ethics that focuses on the consequences of an action. It is summarized by the phrase: do the greatest good for the greatest number of people.

Virtue ethics is an approach to ethics that focuses on a person's character and how actions arise from the presence or absence of various virtues. Commonly held virtues include honesty, compassion or faithfulness.

Sources

FILMOGRAPHY

Bionic Woman (2007) TV series, NBC Universal Television.
Blade Runner (1982) Ridley Scott, The Ladd Company.
Brave New World (1998) Leslie Libman and Larry Williams, Dan Wigutow Productions.
Casino Royale (2006) Martin Campbell, Metro-Goldwyn-Mayer (MGM).
Constant Gardener, The (2005) Fernando Meirelles, Potboiler Productions.
Diving Bell and the Butterfly, The [*Le scaphandre et le papillon*] (2007) Julian Schnabel, Pathé Renn Productions.
DOA: Dead or Alive (2006) Corey Yuen, Constantin Film Produktion.
Fantastic Voyage (1966) Richard Fleischer, Twentieth Century-Fox Film Corporation.
Final Cut, The (2004) Omar Naim, Lions Gate Entertainment.
Gattaca (1997) Andrew Niccol, Columbia Pictures Corporation.
Hear and Now (2007) Irene Taylor Bordsky, Vermillion Films/HBO Documentary Films
I am Legend (2007) Francis Lawrence, Warner Bros. Pictures.
I, Robot (2004) Alex Proyas, Twentieth Century-Fox Film Corporation.
Island, The (2005) Michael Bay, DreamWorks SKG.
John Q (2002) Nick Cassavetes, New Line Cinema.
Logan's Run (1976) Michael Anderson, Metro-Goldwyn-Mayer (MGM).
Matrix, The (1999) Andy Wachowski and Larry Wachowski, Groucho II Film Partnership.
Matrix Reloaded, The (2003) Andy Wachowski and Larry Wachowski, Warner Bros. Pictures.

Matrix Revolutions, The (2003) Andy Wachowski and Larry Wachowski, Warner Bros. Pictures.

Minority Report (2002) Steven Spielberg, Twentieth Century-Fox Film Corporation.

Murder in the Air (1940) Lewis Seiler, Warner Bros. Pictures.

Six Million Dollar Man (1974–8) TV series, Richard Donner and Richard Irving, Harve Bennett Productions.

Slumdog Millionaire (2008) Danny Boyle and Loveleen Tandan, Celador Films.

Spider-Man (2002) Sam Raimi, Columbia Pictures Corporation.

Terminator 3: Rise of the Machines (2003) Jonathan Mostow, C-2 Pictures.

Torn Curtain (1966) Alfred Hitchcock, Universal Pictures.

Total Recall (1990) Paul Verhoeven, Carolco International N.V.

Bibliography

Abraham, C. (2002), 'Gene Pioneer Urges Human Perfection', *Toronto Globe and Mail*, 26 October. www.theglobeandmail.com/servlet/ArticleNews/front/RTGAM/20021026/wxwats1026/Front/homeBN/breakingnews (accessed 6 April 2009).

Adams, T. E. (2008), 'A review of narrative ethics', *Qualitative Inquiry*, 14(2), 175–94.

Adams, W. W. and Baughman, R. H. (2005), 'Richard E. Smalley (1943–2005)', *Science*, 310(5756), 1916.

Agar, N. (2004), *Liberal Eugenics: In Defence of Human Enhancement*. Malden, MA: Blackwell.

AIDS Alert (2008), 'Starpharma optimistic new microbicide will beat odds. VivaGel uses condom coating dosing', *AIDS Alert*, 23(11), 125.

Alda, A. (2005), 'Cybersenses', *Scientific American Frontiers*, 6 April. www.pbs.org/saf/1509/ (accessed 17 March 2009).

Allen, G. (2005), 'The economic promise of nanotechnology', *Issues in Science and Technology*, 12(4), 55–6.

Allhoff, F., Lin, P., Moor J. and Weckert, J. eds (2007), *Nanoethics: The Ethical and Social Implications of Nanotechnology*. Hoboken, NJ: Wiley-Interscientific.

American Friends of Tel Aviv University (2009), 'A fantastic voyage brought to life', 15 January. www.aftau.org/site/News2?page=NewsArticle&id=8465 (accessed 23 January 2009).

Anonymous (1967), 'Control of the asbestos hazard', *Lancet*, 1, 1,311–12.

Anonymous (1992), 'Changing your genes', *The Economist*, April 25, 11–12.

Anonymous (2006), 'Altair Nanotechnologies details long life features of its nano titanate battery', *Altair Nano*, 7 September. www.b2i.us/profiles/investor/ResLibraryView.asp?ResLibraryID=17008&GoTopage=1&BzID=546&Category=856 (accessed 3 January 2008).

Anonymous (2008), 'Nanotechnology paves way for super iPods', *Nanotechwire.com*, 18 April. www.nanotechwire.com/news.asp?nid=5832 (accessed 3 January 2009).

Anonymous (2009), 'Getting to know the public', *Nature Nanotechnology*, 4(2), 71.

Aristotle (1985), *Nicomachean Ethics*. Translated by T. Irwin. Indianapolis, IN: Hackett.

Arrison, S. (2008), 'Science fiction gets funding', *H+*, 1, 32–3.

Ashford, Nicholas, *et al.* (1998), 'Wingspread Statement on the Precautionary Principle'. www.gdrc.org/u-gov/precaution-3.html (accessed 20 April 2009).

Asimov, I. (1966), *Fantastic Voyage*. New York: Bantam Books.

AZoNano, (2008), 'Conducting DNA Tests Quickly and Inexpensively Using New "Lab on a Chip" Technology', *AZoNanotechnology*, 28 September. www.azonano.com/news.asp?newsID=7731 (accessed 25 March 2009).

Bainbridge, W. S. (2003), 'Challenge and response', *TransHumanity*. www.transhumanism.org/index.php/th/more/363/ (accessed 17 April 2009).

Barnes, C. A., Elsaesser, A., Arkusz, J., Smok, A., Palus, J., Lesniak, A., Salvati, A., Hanrahan, J. P., de Jong, W. H., Dziubałtowska, E., Stępnik, M., Rydzyński, K., McKerr, G., Lynch, I., Dawson, K. A. and Howard, C. V. (2008), 'Reproducible Comet Assay of amorphous silica nanoparticles detects no genotoxicity', *Nano Letters*, 8(9), 3,069–74.

Batéjat, D. M. and Lagarde, D. P. (1999), 'Naps and modafinil as countermeasures for the effects of sleep deprivation on cognitive performance', *Aviation, Space, and Environmental Medicine*, 70(5), 493–8.

Bauby, J.-D. (1997), *The Diving Bell and the Butterfly*, translated by J. Leggatt. London: Harper Perennial.

Bawarski, W. E., Chidlowsky, E., Bharali, D. J. and Mousa. S. A. (2008), 'Emerging nanopharmaceuticals', *Nanomedicine*, 4(4), 27382.

Bell, S. (2005), 'From practice research to public policy – the Ministerial Summit on Health Research', *Annals of Pharmacotherapy*, 39(7–8), 1,331–5.

Beltrán, D. J. (2001), 'Preface', in Harremoës *et al.*, pp. 3–5.

Berne, R. W. (2006), *Nanotalk: Conversations with Scientists and Engineers about Ethics, Meaning, and Belief in the Development of Nanotechnology*. Mahwah, NJ: Lawrence Erlbaum.

Berube, D. M. (2006), *Nano-Hype: The Truth Behind the Nanotechnology Buzz*. Amherst, NY: Prometheus Books.

Beveridge and Diamond (2008), 'Consumers Union Requests FDA Safety Assessment on Use of Nanoparticles in Cosmetics and Sunscreens', 11 December. www.bdlaw.com/news-436.html (accessed 5 April 2009).

Bhattachary, D., Stockley, R. and Hunter, A. (2008), *Nanotechnology for Healthcare*. www.epsrc.ac.uk/CMSWeb/Downloads/Other/Report PublicDialogueNanotechHealthcare.pdf (accessed 5 April 2009).

Bhogal, N. and Combes, R. (2007), 'Immunostimulatory antibodies: challenging the drug testing paradigm', *Toxicology In Vitro*, 21(7), 1,227–32.

Bhushan, B. (2006), *Springer Handbook of Nanotechnology* (2nd edn). New York: Springer.

Booth, W. C. (1989), *The Company We Keep: An Ethics of Fiction*. Berkeley, CA: University of California Press.

Booth, W. C. (2001), 'Literary criticism and the pursuit of character', *Literature and Medicine*, 20(2), 97–108.

Bordo, S. (1998), '*Braveheart*, *Babe*, and the contemporary body', in Parens, pp. 189–221.

Borry, P., Schotsmans, P. and Dierickx, K. (2005), 'Developing countries and bioethical research', *New England Journal of Medicine*, 353(8), 852–3.

Bostrom, N. (2004), 'The future of human evolution', in C. Tandy (ed.) *Death and Anti-Death: Two Hundred Years After Kant, Fifty Years After Turing*. Palo Alto, CA: Ria University Press, pp. 339–71.

Bostrom, N. (2005), 'A history of transhumanist thought', *Journal of Evolution and Technology*, 14(1), 1–25.

Bostrom, N. (2008), 'Why I want to be a posthuman when I grow up', in B. Gordijn and R. Chadwick (eds), *Medical Enhancement and Posthumanity*. Berlin: Springer, pp. 107–36.

Bostrom, N. and Roache, R. (2007), 'Human enhancement: ethical issues in human enhancement', in J. Ryberg, T. S. Petersen and C. Wolf (eds), *New Waves in Applied Ethics*. New York: Palgrave Macmillan, pp. 120–52.

Bostrom N., and Sandberg, A. (2009), 'The wisdom of Nature: an evolutionary heuristic for human enhancement', in N. Bostrom and J. Savulescu (eds) *Human Enhancement*. Oxford: Oxford University Press, pp. 375–416.

Boyle, A. (2007), 'Will LiftPort rise again?', *MSNBC*, 18 April. http://cosmiclog.msnbc.msn.com/archive/2007/04/18/157110.aspx (accessed 30 March 2009).

Brake, M. L. and Hook, N. (2008), *Different Engines: How Science Drives Fiction and Fiction Drives Science*. Basingstoke: Macmillan.

Brave, R. (2003), 'James Watson Wants to Build a Better Human', *Center for Genetics and Society*, 28 May. www.geneticsandsociety.org/article.php?id=245 (accessed 6 April 2009).

Brin, D. (1990), 'Afterword', in *Earth*. London: Macdonald, pp. 580–90.

British Museum (n.d.), 'The Lycurgus Cup'. www.britishmuseum.org/explore/highlights/highlight_objects/pe_mla/t/the_lycurgus_cup.aspx (accessed 14 January 2009).

Brock, D. W. (1998), 'Enhancements of human function: some distinctions for policymakers', in Parens, pp. 48–69.

Brooks, D. (2009), 'The End of Philosophy', *New York Times*, 6 April. www.nytimes.com/2009/04/07/opinion/07Brooks.html?_r=1&emc=eta1 (accessed 10 April 2009).

Buchanan, A., Brock, D. W. ,Daniels, N. and Wikler, D. (2000), *From Chance to Choice: Genetics and Justice*. Cambridge: Cambridge University Press.

Buckingham, D. (1987), *Public Secrets: EastEnders and its Audience*. London: BFI.

Caldwell, J. A., Caldwell, J. L., Smith, J. K. and Brown, D. L. (2004), 'Modafinil's effects on simulator performance and mood in pilots during 37 h without sleep', *Aviation, Space, and Environmental Medicine*, 75(9), 777–84.

Card, O. S. (1990), 'Author's Introduction', in *The Worthing Chronicle*. New York: Tor, pp. ix–xiv.

Carlson, E. A. (2001), *The Unfit: A History of a Bad Idea*. Cold Spring Harbor, NY: Cold Spring Harbor Laboratory Press.

Carnes, B. A., Olshansky, S. J. and Grahn, D. (2003), 'Biological evidence for limits to the duration of life', *Biogerontology*, 4(1), 31–45.

Chang, R. S. (2008), 'Tata Nano: The World's Cheapest Car', *New York Times*, 10 January. http://wheels.blogs.nytimes.com/2008/01/10/tata-nano-the-worlds-cheapest-car/ (accessed 28 March 2009).

Charon, R. (2006), *Narrative Medicine: Honouring the Stories of Illness*. Oxford: Oxford University Press.

Chatterjee, A. (2007), 'Cosmetic Neurology and Cosmetic Surgery: Parallels, Predictions and Challenges', *Cambridge Quarterly of Healthcare Ethics*, 16(2), 129–37.

Chirac, P. and Torreele, E. (2006), 'Global framework on essential health R&D', *Lancet*, 367, 1,560–61.

Chorost, M. (2006), *Rebuilt: My Journey Back to the Hearing World*. Boston: Mariner.

Clark, G. M. (2008), 'Personal reflections on the multichannel cochlear implant and a view of the future', *Journal of Rehabilitation Research & Development*, 45(5), 651–94.

Clarke, A. C. (1979), *Fountains of Paradise*. New York: Harcourt Brace Jovanovich.

Clasen, T., Schmidt, W.-P., Rabie, T., Roberts, I. and Cairncross, A. (2007), 'Interventions to improve water quality for preventing diarrhoea: systematic review and meta-analysis', *BMJ*, 334, 782–91.

Cole-Turner, R. (1998), 'Do means matter?' in Parens, pp. 151–61.

Collings, M. (1990), 'Afterword', in O. S. Card, *The Worthing Chronicle*. New York: Tor, pp. 459–63.

Collins, G. P. (2007), 'Shamans of small', *Scientific American Reports*, 17(3), 80–88.

Commission on Health Research for Development (1990), *Health Research: Essential Link to Equity in Development*. New York: Oxford University Press

Conrad, P. (2007), *The Medicalization of Society: On the Transformation of Human Conditions into Treatable Disorders*. Baltimore, MD: Johns Hopkins University Press.

Consumers' Association (2008), *Small Wonder? Nanotechnology and Cosmetics*. www.which.co.uk/documents/pdf/nanotechnology-and-cosmetics-161175.pdf (accessed 5 February 2009).

Court, E. B., Salamanca-Buentello, F., Singer, P. A. and Daar, A. S. (2007), 'Nanotechnology and the developing world', in H. A. M. J. ten Have (ed.) *Nanotechnologies, Ethics and Politics*. Paris: UNESCO, pp. 155–80.

Crichton, M. (2002), *Prey*. London: HarperCollins.

Currall, S. C. (2009), 'Nanotechnology and society: New insights into public perceptions', *Nature Nanotechnology*, 4(2), 79–80.

Currall, S. C., King, E. B., Lane, N., Madera, J. and Turner, S. (2006), 'What drives public acceptance of nanotechnology?' *Nature Nanotechnology*, 1(3), 153–5.

de Grey, A. (2004), 'We will be able to live to 1,000', *BBC News*, 3 December. http://news.bbc.co.uk/2/hi/uk_news/4003063.stm (accessed 20 April 2009).

de Grey, A. D. N. J. (2005), 'Resistance to debate on how to postpone ageing is delaying progress and costing lives', *EMBO Reports*, 6(S1), S49–S53.

Derfus, A. M., Chan, W. C. W. and Bhatia, S. N. (2004), 'Probing the cytotoxicity of semiconductor quantum dots', *Nano Letters*, 4(1), 11–18.

DeSantis, A. D., Webb, E. M. and Noar, S. M. (2008), 'Illicit use of prescription ADHD medications on a college campus: a multimethodological approach', *Journal of American College Health*, 57(3), 315–24.

Dobelle, W. H. (2000), 'Artificial vision for the blind by connecting a television camera to the visual cortex', *ASAIO Journal*, 46(1), 3–9.

Drexler, K. E. (1981), 'Molecular engineering: An approach to the development of general capabilities for molecular manipulation', *Proceedings of the National Academy of Sciences of the USA*, 78(9), 5,275–78.

Drexler, K. E. (1986/2006), *Engines of Creation 2.0: The Coming Era of Nanotechnology* (20th anniversary e-book edition). New York: Anchor Books. www.wowio.com/users/product.asp?BookId=503 (accessed 20 April 2009).

Drexler, K. E. (2001), 'Machine-phase nanotechnology', *Scientific American*, 285(3), 74–5.

Dyson, F. (1997), *Imagined Worlds*. Cambridge, MA: Harvard University Press.

Ebbesen, M. and Jensen, T. G. (2006), 'Nanomedicine: techniques, potentials, and ethical implications', *Journal of Biomedicine and Biotechnology*, 2006(5), 1–11.

Ebert, R. (2005), 'The Island'. http://rogerebert.suntimes.com/apps/pbcs.dll/article?AID=/20050721/REVIEWS/50711003 (accessed 20 October 2008).

Edqvist, L.-E. and Pedersen, K. B. (2001), 'Antimicrobials as growth promoters: resistance to common sense', in Harremoës *et al.*, pp. 93–100.

Edwards, S. A. (2006), *The Nanotech Pioneers: Where Are They Taking Us?* Weinheim: WILEY-VCH.

ElAmin, A. (2006), 'Nanotech database compiles consumer items on the market', *foodproductiondaily.com*, 20 March. www.foodproductiondaily.com/Supply-Chain/Nanotech-database-compiles-consumer-items-on-the-market (accessed 3 January 2009).

Elan (2007), *Technology Focus*. www.elan.com/EDT/nanocrystal_technology/ (accessed 5 April 2009).

Emanuel, E. J., Wendler, D. and Grady, C. (2000), 'What makes clinical research ethical?', *Journal of the American Medical Association*, 283(20), 2701–11.

Erickson, B. E. (2008), 'What's next for nanotechnology?' *Chemical & Engineering News*, 86(32), 35–6.

Erickson, B. E. (2009), 'Nanoceuticals', *Chemical & Engineering News*, 9 February. http://pubs.acs.org/cen/government/87/8706gov3.html (accessed 13 April 2009).

Erickson, M. (2008), 'Small stories and tall tales: Nanotechnology, science fiction and science fact', in A. R. Bell, S. R. Davies and F. Mellor (eds) *Science and its Publics*. Newcastle: Cambridge Scholars Publishing, pp. 135–56.

ETC group (2008), 'Nanotechnology'. www.etcgroup.org/en/issues/nanotechnology.html (accessed 29 December 2008).

European Commission (2000), *Communication from the Commission on the Precautionary Principle*. http://ec.europa.eu/dgs/health_consumer/library/pub/pub07_en.pdf (accessed 16 April 2009).

European Commission (2007a), *Nanosciences and Nanotechnologies: An action plan for Europe 2005–2009. First Implementation Report 2005–2007*. ftp://ftp.cordis.europa.eu/pub/nanotechnology/docs/com_2007_0505_f_en.pdf (accessed 20 April 2009).

European Commission (2007b), *Opinion on Safety of Nanomaterials in Cosmetic Products*. http://ec.europa.eu/health/ph_risk/committees/04_sccp/docs/sccp_o_123.pdf (accessed 5 February 2009).

European Commission (2008), *Code of Conduct for Responsible Nanosciences and Nanotechnologies Research*. ftp://ftp.cordis.europa.eu/pub/fp7/docs/nanocode-recommendation.pdf (accessed 9 March 2008).

European Group on Ethics in Science and New Technologies to the European Commission (2007), *Opinion on the Ethical Aspects in Nanomedicine*. http://ec.europa.eu/european_group_ethics/activities/docs/opinion_21_nano_en.pdf (accessed 7 February 2007).

European Union (1992), *Treaty of Maastricht*. www.eurotreaties.com/maastrichtec.pdf (accessed 19 February 2009).

Expert Scientific Group on Phase One Clinical Trials, (2006), *Final Report*. Norwich: The Stationery Office. www.dh.gov.uk/en/Publicationsand statistics/Publications/PublicationsPolicyAndGuidance/DH_063117 (accessed 16 February 2008).

Farah, M. J. (2008), 'Cognitive enhancement', presentation at the 33rd Annual AAAS Science and Technology Policy Forum, Washington, DC, 8–9 May. www.aaas.org/spp/sfrl/projects/human_enhancement/#ST (accessed 18 March 2009).

Fender, J. K. (2007), 'Patenting trends in nanotechnology', in N. M. de S. Cameron and M. E. Mitchell (eds) *Nanoscale: Issues and Perspectives for the Nano Century*. Hoboken, NJ: Wiley-Interscience, pp. 259–78.

Fernholm, A. (2008), 'Carbon nanotubes may be as harmful as asbestos', *San Francisco Chronicle*, 20 May. www.sfgate.com/cgi-bin/article.cgi?f=/c/a/2008/05/20/BUDG10P518.DTL (accessed 13 February 2009).

Feynman, R. P. (1959/1992), 'There's plenty of room at the bottom', *Journal of Microelectromechanical Systems*, 1(1), 60–6.

FirstScience.tv (2007), *Nano Explorers – Science Fiction or Reality?* www.youtube.com/watch?v=GZqStSP5tKw (accessed 28 December 2008).

Fitzgerald, F. (2000), *Way Out There in the Blue: Reagan, Star Wars and the End of the Cold War*. New York: Simon & Schuster.

Fitzgerald, M. (2009), 'Counting cost of cutting aid', *Irish Times*, 4 April. www.irishtimes.com/newspaper/weekend/2009/0404/12242439911 97.html (accessed 11 April 2009).

Freedman, C. (1998), 'Aspirin for the mind? Some ethical worries about psychopharmacology', in Parens, pp. 135–50.

Freestone, I., Meeks, N., Sax, M. and Higgitt, C. (2007), 'The Lycurgus Cup – A Roman nanotechnology', *Gold Bulletin*, 40(4), 270–7.

Freitas Jr, R. A. (2005), 'Nanotechnology, nanomedicine and nanosurgery', *International Journal of Surgery*, 3(4), 243–6.

Freitas Jr, R. A. (2006), 'Pharmacytes: an ideal vehicle for targeted drug delivery', *Journal of Nanoscience and Nanotechnology*, 6, 2,769–75.

Freitas Jr, R. A. (2007), 'The ideal gene delivery vector: chromallocytes, cell repair nanorobots for chromosome replacement therapy', *Journal of Evolution & Technology*, 16(1), 1–97.

Friedman, E. A. and Kolff, W. J. (2000), 'The beginning of the artificial eye program', *ASAIO Journal*, 46(1), 1–2.

Fukuyama, F. (2004), 'Transhumanism', *Foreign Policy*, 144, 42–3.

Gavaghan, C. (2009), '"No gene for fate?": Luck, harm, and Justice in *Gattaca*', in S. Shapshay (ed.) *Bioethics at the Movies*. Baltimore, MD: Johns Hopkins University Press, pp. 75–86.

Gazit, E. (2007), *Plenty of Room for Biology at the Bottom: An Introduction to Bionanotechnology*. London: Imperial College Press.

Gee, D. and Greenberg, M. (2001), 'Asbestos: from "magic" to malevolent mineral', in Harremoës *et al.*, pp. 52–63.

Gill, M., Haerich, P., Westcott, K., Godenick, K. L. and Tucker, J. A. (2006), 'Cognitive performance following modafinil versus placebo in sleep-deprived emergency physicians: a double-blind randomized crossover study', *Academic Emergency Medicine*, 13(2), 158–65.

Glannon, W. (2002), 'Extending the human life span', *Journal of Medicine and Philosophy*, 27(3), 339–54.

Global Forum for Health Research (2004), *The 10/90 Report on Health Research 2003–2004*. Geneva: Global Forum for Health Research. www.globalforumhealth.org (accessed 21 August 2007).

Gordijn, B. (2007), 'Ethical issues in nanomedicine', in H. A. M. J. ten Have (ed.) *Nanotechnologies, Ethics and Politics*. Paris: UNESCO, pp. 99–123.

Gordon, A. (2004), '*Back to the Future*: Oedipus as time traveller', in S. Redmond (ed.) *Liquid Metal: The Science Fiction Film Reader*. London: Wallflower, pp. 116–25.

Greely, H., Sahakian, B., Harris, J., Kessler, R. C., Gazzaniga, M., Campbell, P. and Farah, M. J. (2008), 'Towards responsible use of cognitive-enhancing drugs by the healthy', *Nature*, 456, 702–5.

Griffitt, R. J., Luo, J., Gao, J., Bonzongo, J.-C. and Barber, D. S. (2008), 'Effects of particle composition and species on toxicity of metallic nanomaterials in aquatic organisms', *Environmental Toxicology and Chemistry*, 27(9), 1,972–8.

Grimes, W. (2008), 'Michael Crichton, author of thrillers, dies at 66', *New York Times*, 5 November. www.nytimes.com/2008/11/06/books/06crichton.html?_r=1 (accessed 3 February 2009).

Grunwald, A. (2005), 'Nanotechnology: a new field of ethical inquiry?' *Science and Engineering Ethics*, 11(2), 187–201.

Habermas, J. (2003), *The Future of Human Nature*. Cambridge: Polity.

Habermas, J. (2006), *Time of Transitions*. Cambridge: Polity.

Hadjipanayis, C. G., Bonder, M. J., Balakrishnan, S., Wang, X., Mao, H. and Hadjipanayis, G. C. (2008), 'Metallic iron nanoparticles for MRI contrast enhancement and local hyperthermia', *Small*, 4(11), 1925–9.

Haldane, J. B. S. (1923), 'Daedalus, or, Science and the future'. www.cscs.umich.edu/~crshalizi/Daedalus.html (accessed 18 March 2009).

Halford, B. (2006), 'The world according to Rick', *Chemical & Engineering News*, 84(41), 13–19.

Hama, R. (2006), 'Couldn't TGN1412 clinical trial tragedy be avoided?' *Informed Prescriber*. www.npojip.org/english/no65.html (accessed 16 February 2008).

Han, S. W., Nakamura, C., Kotobuki, N., Obataya, I., Ohgushi, H., Nagamune, T. and Miyake, J. (2008), 'High-efficiency DNA injection into a single human mesenchymal stem cell using a nanoneedle and atomic force microscopy', *Nanomedicine*, 4(3), 215–25.

Hanson, D. (2008), 'Nanotech strategy', *Chemical & Engineering News*, 86(9), 29–30.

Hansson, S. O. (2004), 'Great Uncertainty About Small Things', *Techné*, 8(2), 26–35.

Harremoës, P., Gee, D., MacGarvin, M., Stirling, A., Keys, J., Wynne, B. and Vaz, S. G. (eds) (2001), *Late Lessons from Early Warnings: The Precautionary Principle 1896–2000*. Copenhagen: European Environment Agency. www.genok.org/filarkiv/File/late_response.pdf (accessed 18 June 2008).

Harris, J. and Holm, S. (2002), 'Extending human lifespan and the precautionary paradox', *Journal of Medicine and Philosophy*, 27(3), 355–68.

Hart, P. D. (2007), *Awareness of and Attitudes Toward Nanotechnology and Federal Regulatory Agencies*. www.nanotechproject.org/process/assets/files/5888/hart_nanopoll_2007.pdf (accessed 11 April 2009).

Hart, P. D. (2008), *Awareness of and Attitudes Toward Nanotechnology and Synthetic Biology*. www.nanotechproject.org/process/assets/files/7040/final-synbioreport.pdf (accessed 20 April 2009).

Harth, W., Seikowski, K., Hermes, B. and Gieler, U. (2008), 'New lifestyle drugs and somatoform disorders in dermatology', *Journal of the European Academy of Dermatology and Venereology*, 22(2), 141–9.

Hassan, M. H. A. (2005), 'Small things and big changes in the developing world', *Science*, 309(5,731), 65–6.

Hawkins, A. H. (2002), 'The idea of character', in R. Charon and M. Montello (eds), *Stories Matter: The Role of Narrative in Medical Ethics*. New York: Routledge, pp. 75–84.

Hawthorne, N. (1843/2003), 'The Birth-Mark', in President's Council on Bioethics, *Being Human*. Washington, DC: President's Council on Bioethics, pp. 5–20. www.bioethics.gov/bookshelf/reader/chapter1.html#reading (accessed 27 January 2007).

Hayflick L. (2007), 'Biological ageing is no longer an unsolved problem', *Annals of the New York Academy of Sciences*, 1,100, 1–13.

Headlam, B. (2000), 'The mind that moves objects', *New York Times Magazine*, 11 June. http://partners.nytimes.com/library/magazine/home/20000611mag-mind.html (accessed 20 April 2009).

Heap, T. (2005), 'Costing the Earth', BBC Radio 4, 14 April. www.bbc.co.uk/radio4/science/costingtheearth_20050414.shtml (accessed 11 April 2009).

Heinlein, R. (1969), *Waldo & Magic, Inc.* London: Pan.

Henderson, M. (2003), 'Let's cure stupidity, says DNA pioneer', *Times Online*, 28 February. www.timesonline.co.uk/tol/news/uk/article1114058.ece (accessed 6 April 2009).

Hickey, H. (2008), 'Contact lenses with circuits, lights a possible platform for superhuman vision', *University of Washington News*, 17 January. http://uwnews.washington.edu/ni/article.asp?articleID=39094 (accessed 5 April 2009).

Highfield, R. (2008), 'Bionic eye heralds cyborg revolution', *Daily Telegraph*, 6 August. www.telegraph.co.uk/scienceandtechnology/science/sciencenews/3348943/Bionic-eye-heralds-cyborg-revolution.html (accessed 17 March 2009).

Hillie, T. and Hlophe, M. (2007), 'Nanotechnology and the challenge of clean water', *Nature Nanotechnology*, 2(11), 663–4.

Hirschfeld, B. (2002), 'Space elevator gets lift', *TechTV Vault*, 31 January. http://g4tv.com/techtvvault/features/35657/Space-Elevator-Gets-Lift.html (accessed 30 March 2009).

Hogg, T. (2007), 'Evaluating microscopic robots for medical diagnosis and treatment', *Nanotechnology Perceptions*, 3(2), 63–73.

Humanity+ (2002), 'The Transhumanist Declaration'. www.transhumanism.org/index.php/WTA/declaration/ (accessed 22 March 2009).

Humanity+ (2009), 'Better than well'. www.transhumanism.org/index.php/WTA/constitution/ (accessed 22 March 2009).

Huxley, A. (1932), *Brave New World*. New York: Bantam.

Hyder, A. A., Akhter, T. and Qayyum, A. (2003), 'Capacity development for health research in Pakistan: the effects of doctoral training', *Health Policy and Planning*, 18(3), 338–43.

IBM (1996), 'Moving atoms'. www.almaden.ibm.com/vis/stm/gallery.html (accessed 16 January 2009).

Iijima, S. and Ichihashi, T. (1993), 'Single-shell carbon nanotubes of 1-nm diameter', *Nature*, 363, 603–5.

Ishiguro, K. (2005), *Never Let Me Go*. London: Faber and Faber.

Jakesevic, N. (2004), 'The world's most dangerous ideas', *Foreign Policy*, 144, 32–3.

John, S. D. (2007), 'How to take deontological concerns seriously in risk–cost–benefit analysis: a re-interpretation of the precautionary principle', *Journal of Medical Ethics*, 33(4), 221–4.

Jömann, N. and Ach, J. S. (2006), 'Ethical implications of nanobiotechnology – state-of-the-art survey of ethical issues related to nanobiotechnology', in J. S. Ach and L. Siep (eds), *Nano-Bio-Ethics: Ethical Dimensions of Nanobiotechnology*. Berlin: Lit Verlag, pp. 13–62.

Jonas, H. (1984), *The Imperative of Responsibility: In Search of an Ethics for the Technological Age*. Chicago: University of Chicago Press.

Juengst, E. (1997), 'Can enhancement be distinguished from prevention in genetic medicine?', *Journal of Medicine and Philosophy*, 22(2), 125–42.

Juengst, E. T. (1998), 'What does *enhancement* mean?' in Parens, pp. 29–47.

Kaiser, J. (2007), 'Gene transfer an unlikely contributor to patient's death', *Science*, 318(5,856), 1,535.

Kass, L. R. (1990), 'Practicing ethics: where's the action?', *Hastings Center Report*, 20(1), 5–12.

katalist (2008), 'Revelation: Tycoon behind Nano Chewing Gum!' *Zimbio*, 25 September. www.zimbio.com/Chewing+gum/articles/144/Revelation+Tycoon+behind+Nano+Chewing+Gum (accessed 3 January 2009).

Kearney, R. (2002), *On Stories*. Abingdon, UK: Routledge.

Kearns, A. J., O'Mathúna, D. P. and Scott, P. A. (2009), 'Diagnostic self-testing: autonomous choices and relational responsibilities', *Bioethics*, Epub.

Khayat, M. H. (n.d.), 'Spirituality in the definition of health: the World Health Organization's point of view'. www.medizin-ethik.ch/publik/spirituality_definition_health.htm (accessed 22 March 2009).

Khushf, G. (2007), 'Upstream ethics in nanomedicine: a call for research', *Nanomedicine*, 2(4), 511–21.

Kleiner, K. (2005), 'Metal: the fuel of the future', *New Scientist*, 2,522, 34.

Ko, H. C., Stoykovich, M. P., Song, J., Malyarchuk, V., Choi, W. M., Yu, C. J., Geddes 3rd, J. B., Xiao, J., Wang, S., Huang, Y. and Rogers, J. A. (2008), 'A hemispherical electronic eye camera based on compressible silicon optoelectronics', *Nature*, 454, 748–53.

Koppe, J. G. and Keys, J. (2001), 'PCBs and the precautionary principle', in Harremoës *et al.*, pp. 64–75.

Kubik, T., Bogunia-Kubik, K. and Sugisaka, M. (2005), 'Nanotechnology on duty in medical applications', *Current Pharmaceutical Biotechnology*, 6(1), 17–33.

Kuhn, T. S. (1962), *The Structure of Scientific Revolutions*. Chicago: University of Chicago Press.

Kumar, R. (2008), 'Approved and investigational uses of modafinil: an evidence-based review', *Drugs*, 68(13), 1803–39.

Lane, N. and Kalil, T. (2005), 'The National Nanotechnology Initiative: present at the creation', *Issues in Science and Technology*, 12(4), 49–54.

Leary, S. P., Liu, C. Y. and Apuzzo, M. L. (2006a), 'Toward the emergence of nanoneurosurgery: part II – nanomedicine: diagnostics and imaging at the nanoscale level', *Neurosurgery*, 58(5), 805–23.

Leary, S. P., Liu, C. Y. and Apuzzo, M. L. (2006b), 'Toward the emergence of nanoneurosurgery: part III – nanomedicine: targeted nanotherapy, nanosurgery, and progress toward the realization of nanoneurosurgery', *Neurosurgery*, 58(6), 1,009–26.

Lem, S. (1973), *The Invincible*. New York: Ace.

Levine, C. and Sugarman, J. (2006), 'After the TGN1412 tragedy: addressing the right questions at the right time for early phase testing', *Bioethics Forum*, 17 April. www.thehastingscenter.org/Bioethicsforum/Post.aspx?id=156 (accessed 7 March 2009).

Levine, C., Faden, R., Grady, C., Hammerschmidt, D., Eckenwiler, L. and Sugarman, J. (2004), '"Special scrutiny": a targeted form of research protocol review', *Annals of Internal Medicine*,140(3), 220–3.

Lin, P. and Allhoff, F. (2007), 'Nanoscience and nanoethics: defining the disciplines', in Allhoff *et al.*, pp. 3–16.

Little, M. O. (1998), 'Cosmetic surgery, suspect norms, and the ethics of complicity', in Parens, pp. 162–76.

Liu, C., Mi, C. C. and Li, B. Q. (2008), 'Energy absorption of gold nanoshells in hyperthermia therapy', *IEEE Transactions on Nanobioscience*, 7(3), 206–14.

Lubick, N. (2008), 'Nanosilver toxicity: ions, nanoparticles – or both?' *Environmental Science & Technology*, 42(23), 8,617.

Lynch, I. and Dawson, K. A. (2008), 'Protein-nanoparticle interactions', *NanoToday*, 3(1–2), 40–7.

MagForce (2009), 'MagForce'. www.magforce.de/english/home1.html (accessed 3 March 2009).

Maher, B. (2008), 'Poll results: look who's doping', *Nature*, 452, 674–5.

Mansoori, G. A. (2005), *Principles of Nanotechnology: Molecular-Based Study of Condensed Matter in Small Systems*. Singapore: World Scientific.

Martuzzi, M. (2007), 'The precautionary principle: in action for public health', *Occupational and Environmental Medicine*, 64(9), 569–70.

McCarthy, J. (2003), 'Principlism or narrative ethics: must we choose between them?', *Medical Humanities*, 29, 65–71.

McGee, E. M. (2008), 'Bioelectronics and implanted devices', in B. Gordijn and R. Chadwick (eds), *Medical Enhancement and Posthumanity*. Berlin: Springer, pp. 207–24.

McGee, E. M. and Maguire, G.Q. (2007), 'Becoming Borg to become immortal: regulating brain implant technologies', *Cambridge Quarterly of Healthcare Ethics*, 16(3), 291–302.

McHugh, J. (2004), 'A Chip in Your Shoulder: Should I get an RFID implant?', *Slate*, 10 November. www.slate.com/id/2109477/ (accessed 10 April 2009).

McKenny, G. P. (1998), 'Enhancements and the ethical significance of vulnerability', in Parens, pp. 222–37.

McKibben, B. (2003), *Enough: Staying Human in an Engineered Age*. New York: Times Books.

Meridian Institute (2005), *Nanotechnology and the Poor: Opportunities and Risks*. www.meridian-nano.org/gdnp/NanoandPoor.pdf (accessed 10 April 2009).

Merkel, R., Boer, G., Fegert, J., Galert, T., Hartmann, D., Nuttin, B. and Rosahl, S. (2007), *Intervening in the Brain: Changing Psyche and Society*. Berlin: Springer.

Merkle, R. C. (1997), 'It's a small, small, small, small world', *Technology Review*, 100(2), 25–32.

Meyer, J. (2008), 'EPA announces regulations for carbon nanotubes and nanoparticles: regulates nanomaterials as chemicals', *On the Edges of*

Science and Law, 5 December. http://blogs.kentlaw.edu/islat/2008/12/epa-announces-regulations-for-carbon-nanotubes-and-nanoparticles-regulates-nanomaterials-as-chemical.html (accessed 12 April 2009).

Milburn, C. (2002), 'Nanotechnology in the age of posthuman engineering: science fiction as science', *Configurations*, 10(2), 261–95.

Miller, H. I. (1994), 'Gene therapy for enhancement', *Lancet*, 344, 317.

Minsky, M. (1986), 'Foreword', in Drexler, pp. 18–21.

Mitchell, C. B., Pellegrino, E. D., Elshtain, J. B., Kliner, J. F. and Rae, S. B. (2007), *Biotechnology and the Human Good*. Washington, DC: Georgetown University Press.

Mnyusiwalla, A., Daar, A. S. and Singer, P. A. (2003), '"Mind the gap": science and ethics in nanotechnology', *Nanotechnology*, 14, R9–R13.

Monthioux, M. and Kuznetsov, V. L. (2006), 'Who should be given the credit for the discovery of carbon nanotubes?', *Carbon*, 44(9), 1,621–3.

Myhr, A. I. and Dalmo, R. A. (2007), 'Nanotechnology and risk: what are the issues?', in Allhoff *et al.*, pp. 149–59.

Naam, R. (2008), 'The distribution of post-humanity', *H+*, 1, 23.

NanoEthics Group (2003–2008), 'What's the big deal?' www.nanoethics.org/bigdeal.html (accessed 14 February 2007).

Nanooze (2005), 'Q&A: Don Eigler', *Nanooze*. www.nanooze.org/english/interviews/doneigler.html (accessed 15 January 2009).

Nanowerk (2007), 'Debunking the trillion dollar nanotechnology market size hype', *Nanowerk Spotlight*, 18 April. www.nanowerk.com/spotlight/spotid=1792.php (accessed 12 April 2009).

NASA (2008), 'Extension of the human senses'. www.nasa.gov/centers/ames/research/technology-onepagers/human_senses.html (accessed 15 March 2009).

National Science Foundation (2001), *Societal Implications of Nanoscience and Nanotechnology*. www.wtec.org/loyola/nano/NSET.Societal.Implications/nanosi.pdf (accessed 20 April 2009).

NC3Rs (n.d.). 'The 3Rs'. www.nc3rs.org.uk/page.asp?id=7 (accessed 5 April 2009).

NCI (2006), *Questions and Answers: Centers of Cancer Nanotechnology Excellence.* http://nano.cancer.gov/about_alliance/CCNE_QA.pdf (accessed 5 March 2009).

NNI (2003), *National Nanotechnology Initiate: Research and Development Supporting the Next Industrial Revolution. Supplement to the President's*

FY 2004 Budget, Washington, DC: Office of Science and Technology Policy. www.nano.gov/nni04_budget_supplement.pdf (accessed 10 April 2009).

NNI (2008a), *Strategy for Nanotechnology-Related Environmental, Health, and Safety Research*, Washington, DC: Office of Science and Technology Policy. www.nano.gov/NNI_EHS_Research_Strategy.pdf (accessed 14 February 2009).

NNI (2008b), *Research and Development Leading to a Revolution in Technology and Industry. Supplement to the President's 2009 Budget*, Washington, DC: Office of Science and Technology Policy. www.nano.gov/NNI_08Budget.pdf (accessed 16 January 2009).

Nordmann, A. (2007), 'If and then: a critique of speculative nanoethics', *NanoEthics*, 1(1), 31–46.

NRC (2006), *A Matter of Size: Triennial Review of the National Nanotechnology Initiative*. www.nap.edu/catalog/11752.html (accessed 20 April 2009).

Nussbaum, M. C. (2001), *Upheavals of Thought: The Intelligence of Emotions*. Cambridge: Cambridge University Press.

O'Connor, A. (2008), 'Tata Nano – world's cheapest new car is unveiled in India', *The Times*, 11 January. www.timesonline.co.uk/tol/driving/article3164205.ece (accessed 28 March 2009).

O'Mathúna, D. P. (1997), 'The case of human growth hormone', in J. F. Kilner, R. D. Pentz and F. E. Young (eds), *Genetics and Ethics: Do the Ends Justify the Genes?*, Grand Rapids, MI: Eerdmans, pp. 203–17.

O'Mathúna, D. P. (2002), 'Genetic technology, enhancement, and Christian values', *National Catholic Bioethics Quarterly*, 2(2), 277–95.

O'Mathúna, D. P. (2006), 'Human dignity in the Nazi Era: implications for current bioethics', *BMC Medical Ethics*, 7, 2.

O'Mathúna, D. P. (2007a), 'Ethical issues with healthcare research in developing countries', *Research Practitioner*, 8(3), 92–100.

O'Mathúna, D. P. (2007b), 'Decision-making and health research: ethics and the 10/90 gap', *Research Practitioner*, 8(5), 164–72.

O'Mathúna, D. P. (2007c), 'Bioethics and biotechnology', *Cytotechnology*, 53(1–3), 113–19.

O'Mathúna, D. P. and McAuley, A. (2006), *Counterfeit Drugs: Towards an Irish Response to a Global Crisis*. Dublin: Irish Patients' Association. www.dcu.ie/nursing/counterfeit_drugs.shtml (accessed 25 August 2007).

O'Neill, O. (2002), *Autonomy and Trust in Bioethics*. Cambridge: Cambridge University Press.

Obataya, I., Nakamura, C., Han, S. W., Nakamura, N. and Miyake, J. (2005), 'Nanoscale operation of a living cell using an atomic force microscope with a nanoneedle', *Nano Letters*, 5(1), 27–30.

Oberdörster, E. (2004), 'Manufactured nanomaterials (fullerenes, C60) induce oxidative stress in brain of juvenile largemouth bass', *Environmental Health Perspectives*, 112(10), 1,058–62.

Oberdörster, G., Oberdörster, E. and Oberdörster, J. (2005), 'Nanotoxicology: an emerging discipline evolving from studies of ultrafine particles', *Environmental Health Perspectives*, 113(7), 823–39.

OECD (2005), *Opportunities and Risks of Nanotechnologies*. www.oecd.org/dataoecd/37/19/37770473.pdf (accessed 7 February 2009).

Olshansky, S. J. (2008), 'The science of ageing and life extension', presentation at Extending Life: Setting the Agenda for the Ethics of Ageing, Death & Immortality, Phoenix, AZ, 6–8 March.

Olshansky, S. J., Hayflick, L. and Carnes, B. A. (2002a), 'No truth to the fountain of youth', *Scientific American*, 286(6), 92–5.

Olshansky, S. J., Hayflick, L. and Carnes, B. A. (2002b), 'Position statement on human ageing', *Journals of Gerontology. Series A, Biological Sciences and Medical Sciences*, 57(8), B292–97.

Olshansky, S. J., Carnes, B. A. and Butler, R. N. (2003), 'If humans were built to last', *Scientific American Special Editions*, 13(2), 94–100.

Olshansky, S. J., Butler, R. N. and Carnes, B. A. (2007), 'What if humans were designed to last?' *The Scientist*, 21(3), 28–35.

Orwell, G. (1949), *Nineteen Eighty-Four*. London: Secker and Warburg.

Ovid. 'Daedalus and Icarus', *Metamorphoses*, Book VIII, 183–235.

Ozin, G. A., Arsenault, A. C. and Cademartiri, L. (2009), *Nanochemistry: A Chemical Approach to Nanomaterials* (2nd edn). Cambridge: Royal Society of Chemistry.

Pang, T., Pablos-Mendez, A. and Usselmuiden, C. (2004), 'From Bangkok to Mexico: towards a framework for turning knowledge into action to improve health systems', *Bulletin of the World Health Organization*, 82(10), 720–2.

Parens, E. (ed). (1998), *Enhancing Human Traits: Ethical and Social Implications*. Washington, DC: Georgetown University Press.

Pascal, B. (1660), *Pensées*, trans. W. F. Trotter. www.leaderu.com/cyber/books/pensees/pensees.html (accessed 19 April 2009).

Peer, D., Park, E. J., Morishita, Y., Carman, C. V. and Shimaoka, M. (2008), 'Systemic leukocyte-directed siRNA delivery revealing cyclin D1 as an anti-inflammatory target', *Science*, 319(5863), 627–30.

Peterson, C. and Heller, J. (2007), 'Nanotech's promise: overcoming humanity's more pressing challenges', in Allhoff *et al.*, pp. 57–70.

Peto, J. (1999), 'The European mesothelioma epidemic', *British Journal of Cancer*, 79(3-4), 666–72.

Phoenix, C. (2003), 'Don't let Crichton's Prey scare you – the science isn't real', *Nanotechnology Now*. www.nanotech-now.com/Chris-Phoenix/prey-critique.htm (accessed 3 February 2009).

Phoenix, C. and Treder, J. (2004), 'Applying the precautionary principle to nanotechnology', *Center for Responsible Nanotechnology*. http://crnano.org/precautionary.htm (accessed 6 March 2008).

Picoult, J. (2004), *My Sister's Keeper*. New York: Atria Books.

Pitt, J. (2006), 'When is an image not an image?', in J. Schummer and D. Baird (eds), *Nanotechnology Challenges: Implications for Philosophy, Ethics and Society*. Singapore: World Scientific, pp. 131–41.

Pogge, T. W. (2005a), 'Human rights and global health: a research program', *Metaphilosophy*, 36, 182–209.

Pogge, T. W. (2005b), 'Real world justice', *Journal of Ethics*, 9, 29–53.

Pogge, T. (2008), *World Poverty and Human Rights* (2nd edn). Cambridge: Polity.

Poland, C. A., Duffin, R., Kinloch, I., Maynard, A., Wallace, W. A. H., Seaton, A., Stone, V., Brown, S., MacNee, W. and Donaldson, K. (2008), 'Carbon nanotubes introduced into the abdominal cavity of mice show asbestos-like pathogenicity in a pilot study', *Nature Nanotechnology*, 3(7), 423–8.

Postman, N. (1992), *Technopoly: The Surrender of Culture to Technology*. London: Vintage.

President's Council on Bioethics (2003a), *Being Human*. Washington, DC: President's Council on Bioethics.

President's Council on Bioethics (2003b), *Beyond Therapy: Biotechnology and the Pursuit of Happiness*. Washington, DC: President's Council on Bioethics. www.bioethics.gov/reports/beyondtherapy/ (accessed 20 April 2009).

Project on Emerging Nanotechnologies (2009), 'Consumer Products', www.nanotechproject.org/inventories/consumer/ (accessed 20 April 2009).

Prow, T. W., Bhutto, I., Kim, S. Y., Grebe, R., Merges, C., McLeod, D. S., Uno, K., Mennon, M., Rodriguez, L., Leong, K. and Lutty, G. A. (2008), 'Ocular nanoparticle toxicity and transfection of the retina and retinal pigment epithelium', *Nanomedicine: Nanotechnology, Biology and Medicine*, 4(4), 340–49.

Qin, Y., Wang, X. and Wang, Z. L. (2008), 'Microfibre–nanowire hybrid structure for energy scavenging', *Nature*, 451, 809–13.

Ramsay, S. (2001), 'No closure in sight for the 10/90 health-research gap', *Lancet*, 358, 1348.

Ray, P. C., Yu, H. and Fu, P. P. (2009), 'Toxicity and environmental risks of nanomaterials: Challenges and future needs', *Journal of Environmental Science and Health. Part C. Environmental Carcinogenesis & Ecotoxicology Reviews*, 27(1), 1–35.

Redmond, S. (2004), 'The origin of the species: time travel and the primal scene', in S. Redmond (ed.), *Liquid Metal: The Science Fiction Film Reader*. London: Wallflower, pp. 114–15.

Regis, E. (1990), *Great Mambo Chicken and the Transhuman Condition: Science Slightly Over the Edge*. London: Penguin.

Reith, M. (2003), *Nano-Engineering in Science and Technology: An Introduction to the World of Nano-Design*. River Edge, NJ: World Scientific.

Rennie, S. and Mupenda, B. (2008), 'Living apart together: reflections on bioethics, global inequality and social justice', *Philosophy, Ethics, and Humanities in Medicine*, 3, 25.

Resnik, D. B. and Tinkle, S. S. (2007a), 'Ethics in nanomedicine', *Nanomedicine*, 2(3), 345–50.

Rethorst, J. (1997), 'Art and imagination, implications of cognitive science for moral education', *Philosophy of Education*. www.ed.uiuc.edu/EPS/PES-yearbook/97_docs/rethorst.html (accessed 2 April 2009)

Reynolds, A. (2001), *Chasm City*. London: Gollancz.

Reynolds, G. H. (2002), 'Falling *Prey* to science fiction', *TCS Daily*, 25 November. www.techcentralstation.com/ (accessed 3 February 2009).

Reynolds, G. H. (2004), 'A tale of two nanotechs', *TCS Daily*, 28 January. www.techcentralstation.com/012804A.html (accessed 30 March 2009).

Robinson, J. (2006), *Deeper than Reason: Emotion and its Role in Literature, Music, and Art*. Oxford: Oxford University Press.

Roco, M. C. and Bainbridge, W. S. (eds). (2003), *Converging Technologies for Improving Human Performance: Nanotechnology, Biotechnology, Information Technology and Cognitive Science*. Dordrecht: Kluwer Academic.

Roukes, M. (2007), 'Plenty of room indeed', *Scientific American Reports*, 17(3), 4–11.

Royal Academy of Engineering (2009), *Synthetic Biology: Scope, Applications and Implications*, 6 May. www.raeng.org.uk/news/publications/list/reports/Synthetic_biology.pdf (accessed 12 May 2009).

Royal Society and Royal Academy of Engineering (2004), *Nanoscience and Nanotechnologies: Opportunities and Uncertainties*. www.nanotec.org.uk/finalReport.htm (Accessed 22 December 2006.)

Royal Statistical Society (2007), *Report of the Working Party on Statistical Issues in First-in-Man Studies*. www.rss.org.uk/main.asp?page=2816 (accessed February 23, 2008).

Rubin, C. T. (2005), 'Daedalus and Icarus revisited', *New Atlantis*, 8, 73–91.

Rumsfeld, D. H. (2002), 'Department of Defense News Briefing', 12 February. www.defenselink.mil/Transcripts/Transcript.aspx?TranscriptID=2636 (accessed 4 April 2009).

Russell, B. (1924), 'Icarus, or, The future of science'. www.cscs.umich.edu/~crshalizi/Icarus.html (accessed 18 March 2009).

Sabin, J. E. and Daniels, N. (1994), 'Determining "Medical Necessity" in Mental Health Practice', *Hastings Center Report*, 24(6), 5–13.

Sandel, M. (2004), 'The case against perfection', *Atlantic Monthly*, 293(3), 50–62.

Sandel, M. (2007), *The Case Against Perfection: Ethics in the Age of Genetic Engineering*. Cambridge, MA: Harvard University Press.

Sanhai, W. R., Sakamoto, J. H., Canady, R. and Ferrari, M. (2008), 'Seven challenges for nanomedicine', *Nature Nanotechnology*, 3(5), 242–4.

Sawisch, L. P. (1986), 'Psychosocial aspects of short stature: the day to day context', in B. Stabler and L. E. Underwood (eds), *Slow Grows the Child: Psychosocial Aspects of Growth Delay*. Hillsdale, NJ: Lawrence Erlbaum Associates, pp. 46–56.

SCENIHR: Scientific Committee on Emerging and Newly Identified Health Risks (2009), *Risk Assessment of Products of Nanotechnologies*. http://ec.europa.eu/health/ph_risk/committees/04_scenihr/docs/scenihr_o_023.pdf (accessed 2 March 2009).

Schettler, T. and Raffensperger, C. (2004), 'Why is a precautionary approach needed?', in M. Martuzzi and J. A. Tickner (eds), *The Precautionary Principle: Protecting Public Health, the Environment and the Future of our Children*. Copenhagen: WHO, pp. 63–84. www.euro.who.int/document/e83079.pdf (accessed 20 April 2009).

Scheufele, D. A., Corley, E. A., Dunwoody, S., Shih, T.-J., Hillback, E. and Guston, D. H. (2007), 'Scientists worry about some risks more than the public', *Nature Nanotechnology*, 2(12), 732–4.

Schulz, W. (2000), 'Crafting a national nanotechnology effort', *Chemical & Engineering News*. http://pubs.acs.org/cen/nanotechnology/7842/7842government.html (accessed 22 January 2009).

Schummer, J. (2007a), 'Identifying ethical issues of nanotechnologies', in H. A. M. J. ten Have (ed.), *Nanotechnologies, Ethics and Politics*. Paris: UNESCO, pp. 79–98.

Schummer, J. (2007b), 'Impact of nanotechnologies on developing countries', in Allhoff *et al.*, pp. 291–307.

Shapshay, S. (ed). (2009a), *Bioethics at the Movies*. Baltimore, MD: Johns Hopkins University Press.

Shapshay, S. (2009b), 'Lifting the genetic veil of ignorance: is there anything really unjust about Gattacan society?' in S. Shapshay (ed.), *Bioethics at the Movies*. Baltimore, MD: Johns Hopkins University Press, pp. 87–101.

Sheldon S. and Wilkinson S. (2004), 'Should selecting saviour siblings be banned?' *Journal of Medical Ethics*, 30(6), 533–7.

Shelley, M. (2003), *Frankenstein*. London: Penguin.

Singer, P. (1972), 'Famine, affluence and morality', *Philosophy and Public Affairs*, 1(3), 229–43.

Sirius, R. U. and Duncan, D. E. (2008), 'Manipulating evolution', *H+*, 1, 13.

Small Times (2004), 'As nanotech grows, leaders grapple with public fear and misperception', *Small Times*, 20 May. www.smalltimes.com/Articles/Article_Display.cfm?ARTICLE_ID=269457&p=109 (accessed 9 February 2009).

Smalley, R. E. (2001), 'Of chemistry, love and nanobots', *Scientific American*, 285(3), 76–7.

Smith, M. M. (1996), *Spares: Ever Thought of Cloning Yourself?* London: HarperCollins.

Smith, R. (2008), 'Blind see again thanks to bionic eye', *Daily Telegraph*,

21 April 2008, www.telegraph.co.uk/news/1895441/Blind-see-again-thanks-to-bionic-eye.html (accessed 17 March 2009).

Smith-Coggins, R., Howard, S. K., Mac, D. T., Wang, C., Kwan, S., Rosekind, M. R., Sowb, Y., Balise, R., Levis, J. and Gaba, D. M. (2006), 'Improving alertness and performance in emergency department physicians and nurses: the use of planned naps', *Annals of Emergency Medicine*, 48(5), 596–604.

Someya, T. (2008), 'Electronic eyeballs', *Nature*, 454, 703–4.

Spaceward Foundation (2008), 'Elevator:2010 – The Space Elevator Challenge'. www.spaceward.org/elevator2010 (accessed 12 April 2009).

Sreenivasan, G. and Benatar, S. R. (2006), 'Challenges for global health in the 21st century: some upstream considerations', *Theoretical Medicine and Bioethics*, 27(1), 3–11.

Staal, F. J., Pike-Overzet, K., Ng, Y. Y. and van Dongen, J. J. (2008), 'Sola dosis facit venenum. Leukemia in gene therapy trials: a question of vectors, inserts and dosage?', *Leukemia*, 22(10), 1,849–52.

Stephenson, N. (1995), *The Diamond Age*. London: Penguin.

Stevens, P. (2004), *Diseases of Poverty and the 10/90 Gap*. London: International Policy Network. www.fightingdiseases.org/pdf/Diseases_of_Poverty_FINAL.pdf (accessed 22 August 2007).

Steyn, M. (2003), 'Rummy speaks the truth, not gobbledygook', *Telegraph*, 8 December. www.telegraph.co.uk/comment/personal-view/3599959/Rummy-speaks-the-truth-not-gobbledygook.html (accessed 13 April 2009).

Sun, C., Lee, J. S. and Zhang, M. (2008), 'Magnetic nanoparticles in MR imaging and drug delivery', *Advanced Drug Delivery Reviews*, 60(11), 1,252–65.

Suntharalingam, G., Perry, M. R., Ward, S., Brett, S. J., Castello-Cortes, A., Brunner, M. D. and Panoskaltsis, N. (2006), 'Cytokine storm in a phase 1 trial of the anti-CD28 monoclonal antibody TGN1412', *New England Journal of Medicine*, 355(10), 1,018–28.

ten Have, H. A. M. J. (2007), 'Introduction', in H. A. M. J. ten Have (ed.), *Nanotechnologies, Ethics and Politics*. Paris: UNESCO, pp. 13–35.

Thomas, T. and Foster, G. (2007), 'Nanomedicines in the treatment of chronic hepatitis C – focus on pegylated interferon alpha-2a', *International Journal of Nanomedicine*, 2(1), 19–24.

Tropical Disease Research (2005), *Making Health Research Work for Poor People: Progress 2003–2004*. Geneva: World Health Organization.

Trouiller, P., Olliaro, P., Torreele, E., Orbinski, J., Laing, R. and Ford, N. (2002), 'Drug development for neglected diseases: a deficient market and a public-health policy failure', *Lancet*, 359, 2,188–94.

Turner, L. (2004), 'Bioethics needs to rethink its agenda', *BMJ*, 328, 175.

UNESCO (2005a), *The Precautionary Principle: World Commission on the Ethics of Scientific Knowledge and Technology (COMEST)*. Paris: UNESCO. http://unesdoc.unesco.org/images/0013/001395/139578e.pdf (accessed 19 April 2009).

UNESCO (2006), *The Ethics and Politics of Nanotechnology*. Paris: United Nations Educational, Scientific and Cultural Organization. http://unesdoc.unesco.org/images/0014/001459/145951e.pdf (accessed 6 March 2008).

United Nations (1992), *Rio Declaration on Environment and Development*. www.unep.org/Documents.Multilingual/Default.asp?documentid=78&articleid=1163 (accessed 19 April 2009).

UPI (2005), 'Apple shaken by iPod Nano lawsuit', *Physorg.com*, 24 October. www.physorg.com/news7522.html (accessed 3 January 2009).

Vidyasagar, D. (2006), 'Global notes: the 10/90 gap disparities in global health research', *Journal of Perinatology*, 26(1), 55–6.

Viktor, E. (2005), 'Spaceworld2000'. www.spaceworld2000.com/ (accessed 29 March 2009).

von Bubnoff, A. (2006), 'Study shows no nano in Magic Nano, the German product recalled for causing breathing problems', *Small Times*, 26 May. www.smalltimes.com/Articles/Article_Display.cfm?ARTICLE_ID=270664&p=109 (accessed 3 April 2009).

von Kraus, M. K. and Harremoës, P. (2001), 'MTBE in petrol as a substitute for lead', in Harremoës *et al.*, pp. 110–25.

Wagner, V., Dullaart, A., Bock, A. and Zweck, A. (2006), 'The emerging nanomedicine landscape', *Nature Biotechnology*, 24, 1,211–17.

Walker, J. (1990), 'Writing with atoms'. www.fourmilab.ch/autofile/www/section2_84_14.html (accessed 23 January 2007).

Walton, D. N. (1989), *Informal Logic: A Handbook for Critical Argumentation*. Cambridge: Cambridge University Press.

Warner, H., Anderson, J., Austad, S., Bergamini, E., Bredesen, D., Butler, R., Carnes, B. A., Clark, B. F., Cristofalo, V., Faulkner, J., Guarente, L., Harrison, D. E., Kirkwood, T., Lithgow, G., Martin, G., Masoro, E., Melov, S., Miller, R. A., Olshansky, S. J., Partridge, L., Pereira-Smith, O., Perls, T., Richardson,

A., Smith, J., von Zglinicki, T., Wang, E., Wei, J. Y. and Williams, T. F. (2005), 'Science fact and the SENS agenda. What can we reasonably expect from ageing research?' *EMBO Reports*, 6(11), 1,006–1,008.

Weckert, J. and Moor, J. (2007), 'The precautionary principle in nanotechnology', in Allhoff *et al.*, pp. 133–46.

Wei, F., Wang, G.-D., Kerchner, G. A., Kim, S. J., Xu, H.-M., Chen, Z.-F. and Zhuo, M . (2001), 'Genetic enhancement of inflammatory pain by forebrain NR2B overexpression', *Nature Neuroscience*, 4(2), 164–9.

Wells, H. G. (1914), *The World Set Free*. Quill Pen Classics.

Wells, H. G. (1984), *The Time Machine and The Invisible Man*. New York: Penguin.

Wen, J. Legendre, L. A. Bienvenue, J. M. and Landers, J. P. (2008), 'Purification of nucleic acids in microfluidic devices', *Analytical Chemistry*, 80(17), 6472–9.

WHO (1948), *Constitution*. www.who.int/about/definition/en/print.html (accessed 20 April 2009).

WHO (2004a), 'Dealing with uncertainty – how can the precautionary principle help protect the future of our children?' in M. Martuzzi and J. A. Tickner (eds), *The Precautionary Principle: Protecting Public Health, the Environment and the Future of our Children*. Copenhagen: WHO, pp. 15–30. www.euro.who.int/document/e83079.pdf (accessed 20 April 2009).

WHO (2004b), *The World Health Report 2004: Changing History*. Geneva: World Health Organization. www.who.int/whr/2004/en/index.html (accessed 22 August 2007).

Wilson, J. M. (2009), 'Lessons learned from the gene therapy trial for ornithine transcarbamylase deficiency', *Molecular Genetics and Metabolism*, 96(4), 151–7.

Wolfe, J. (2006), 'Top nano products of 2005', *Forbes*, 10 January. www.forbes.com/2006/01/10/apple-nano-in_jw_0109soapbox.inl.html (accessed 10 April 2009).

Wolfe, J. (2009), 'Five technologies set to change the decade', *Forbes*, 7 January. www.forbes.com/2009/01/07/firstsolar-ibm-cambrios-leader ship-clayton-in-cx_jw_0106claytonchristensen_inl.html (accessed 10 April 2009).

Worldwatch Institute (2000), *Chronic Hunger and Obesity Epidemic; Eroding Global Progress*. www.worldwatch.org/node/1672 (accessed 11 April 2009).

Wright, D. and Gronlund, L. (2008), 'Twenty-five years after Reagan's Star Wars speech', *Bulletin of the Atomic Scientists*. www.thebulletin.org/web-edition/features/twenty-five-years-after-reagans-star-wars-speech (accessed 20 January 2009).

Yates, W. (1988), 'Literature, the arts, and the teaching of ethics: the survey', in D. M. Yeager (ed.), *Annual of the Society of Christian Ethics*. Washington: Georgetown University Press, pp. 225–37.

Zhao, Y., Xing, G. and Chai, Z. (2008), 'Are carbon nanotubes safe?' *Nature Nanotechnology*, 3(4), 191–2.

Žižek, S. (2004), 'What Rumsfeld doesn't know that he knows about Abu Ghraib', *In These Times*, 21 May. www.lacan.com/zizekrumsfeld.htm (accessed 13 April 2009).

Index

Bold page numbers refer to Glossary entries.